Inquiry-Based Laboratories for Liberal Arts Chemistry

Vickie M. Williamson
Texas A&M University

M. Larry Peck
Texas A&M University

Australia • Brazil • Canada • Mexico • Singapore • Spain • United Kingdom • United States

Printed in the United States of America
2 3 4 5 6 7 10 09 08 07 06

Printer: ePAC

0-495-01515-6
Cover Image: George B. Diebold/Corbis

Thomson Higher Education
10 Davis Drive
Belmont, CA 94002-3098
USA

Contents

Foreword

TO: STUDENTS AND INSTRUCTORS

This laboratory manual is designed for chemistry courses for non-science majors. The focus of the manual is conceptual learning of the chemical phenomena in our lives. The manual employs the learning cycle approach, which is used as the underlying model for the guided and open inquiry/application laboratories. The learning cycle is derived from learning theory, is consistent with the nature of science, and has three sequential phases: 1) exploring/gathering data; 2) discussion/concept invention; 3) expansion/application. The titles of these phases have changed with various curricula. During the exploration phase, students are actively involved with materials in experimentation and gathering data. The concept invention phase is an inductive activity involving logical organization, comparison, and interpreting of data, resulting in a generalization about the variables. During the application phase, students are asked to apply the generalization in a new situation or examine another aspect of the concept. In this deductive phase, an application or open inquiry laboratory could be used.

A guided inquiry laboratory is used to accomplish the exploring and concept invention phases of the learning cycle. In the Procedures section of a guided inquiry laboratory, students are guided to collect data on a number of variables. Then in the Analysis section students are asked to describe any patterns between the variables. Students can apply various mathematical operations (+, –, /, ×) and/or use graphing. The Concept Questions of the guided inquiry laboratories allow students to give their understandings of the relationships *before* the student text or instructor describes the concept. Students establish that a relationship exists before the scientific terminology is applied to the relationship. After the concept has been developed, students apply and expand the concept in Application Questions, an Application laboratory, or an Open Inquiry laboratory, in which students design their own procedures to answer questions, which require that they apply the concept.

We feel that use of the learning cycle model is more like the procedures employed by actual scientists. From observations, scientists inductively form generalizations (hypotheses). These generalizations are then deductively applied to other specific situations to see if the generalization holds. The learning cycle uses an inductive-deductive cycle. This varies from the traditional mode of teaching where students are told the generalization then use the laboratory to verify that the generalization is in fact correct, using only deductive thinking. Additionally, educational research shows that deeper learning occurs with the learning cycle.

The role of the instructor varies with each type of experiment. The goals of the guided inquiry laboratories are to direct the students to collect data on variables without previously studying the concepts and to guide the students to look for patterns in the data, so that students can form their own conceptual learning about the relationship between the variables. The goal of the application/open inquiry laboratories is to allow the students to apply a concept or relationship in a new setting. While application laboratories give more direction, they still require the application of a previous learned concept and allow the students to design some part of their procedures. They allow the instructor to judge whether a student can transfer learning. Skills are also developed in both guided inquiry and application or open inquiry laboratories.

Vickie M. Williamson and M. Larry Peck

Laboratory Safety

A chemical laboratory can be a hazardous place to work if basic safety rules are not enforced. If the safety rules are strictly enforced, the chance of one being injured becomes very small. In this course experiments that involve some of the safer chemical and equipment have been selected. However you will still need to use electric equipment, hot water, and concentrated solutions. These are only safe to work with if you follow the correct procedures. With proper understanding of what you are doing, careful attention to safety precautions, and adequate supervision, you will find the chemical laboratory to be a safe place in which you can learn much about chemistry.

Laboratory accidents belong to two general categories of undesirable events: mishaps caused by your own negligence and accidents beyond your control. Although accidents in the laboratory fortunately are rather rare events, you nevertheless must be familiar with all safety rules and emergency procedures. If you know and follow safe working practices, you will pose no threat of serious harm to yourself or others.

Every laboratory system will have some unique features for the prevention of accidents or for handling emergencies. It is important that you become thoroughly familiar with the special safety aspects of your own laboratory area. Some general precautions and procedures applicable to any chemical laboratory are summarized below.

SAFETY

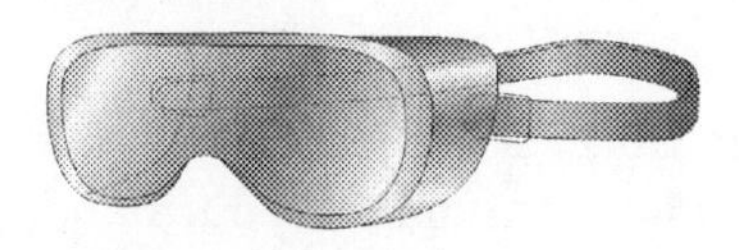

Safety goggles

1. **WEAR APPROVED EYE PROTECTION AT ALL TIMES.** Very minor laboratory accidents, such as the splattering of solution can cause permanent eye damage. Wearing laboratory goggles can prevent this eye damage. In the chemistry teaching laboratories safety goggles of an **approved** type **must be worn** by all persons in the room at all times that anyone is working with or transporting glassware or conducting any experimental work. Experimental work includes simple tasks such as transporting chemicals or glassware, obtaining quantitative measurements that involve non-sealed containers, etc. Lightweight "visitors' shields" or prescription glasses with side shields are acceptable only for laboratory visitors, but are **not** suitable for routine laboratory work.

2. **WEAR PROPER PROTECTIVE CLOTHING.** Proper protective clothing **must** be worn by all persons in the room at all times that anyone is working with or transporting glassware or conducting any experimental work. Exposed skin is particularly susceptible to injury by splattering of hot, caustic, or flammable materials. Students and instructors need to be protected from their necks to below their knees. This requirement includes **no shorts, no short skirts, no sleeveless garments,** and **no bare midriffs**. Long lab coats or aprons are required if shorts or short skirts are worn. Makeshift coverage such as shirts being used as aprons, paper taped over the knees, etc., is not considered to be suitable. Tight fitting clothing, long unrestrained hair, clothing that contains excessive fringe or even overly loose-fitting clothing may be ruled to be unsafe.

3. **WEAR PROPER PROTECTIVE FOOTWEAR.** No sandals, no open-toed shoes, and no foot covering with absorbent soles are allowed. Any foot protection that exposes any part of one's toes is unsuitable for wear in the laboratory.

4. **NEVER EAT, DRINK, OR SMOKE IN A CHEMICAL LABORATORY.** Tiny amounts of some chemicals may cause toxic reactions. Many solvents are easily ignited. Food and drinks are **never** allowed in the labs. This includes all visible insulated water bottles or mugs, containers of water or flavored drinks, containers of ice intended for consumption, etc. If a food or drink container is empty or unopened, it needs to be inside a backpack, etc., and out of sight.

5. **NEVER WORK IN A CHEMICAL LABORATORY WITHOUT PROPER SUPERVISION.** Your best protection against accidents is the presence of a trained, conscientious supervisor, who is watching for potentially dangerous situations and who is capable of properly handling an emergency.

6. **NEVER PERFORM AN UNAUTHORIZED EXPERIMENT.** "Simple" chemicals may produce undesired results when mixed. Any experimentation not requested by the laboratory manual or approved by your instructor may be considered to be unauthorized experimentation.

7. **NEVER INHALE GASES OR VAPORS UNLESS DIRECTED TO DO SO.** If you must sample the odor of a gas or vapor, use your hand to waft a small sample toward your nose.

8. **EXERCISE PROPER CARE IN HEATING OR MIXING CHEMICALS.** Be sure of the safety aspects of every situation in advance. For example, never heat a liquid in a test tube that is pointed toward you or another student. Never pour water into a concentrated acid. Proper dilution technique requires that the concentrated reagent be slowly poured into water while you stir to avoid localized overheating.

9. **BE CAREFUL WITH GLASS EQUIPMENT.** Cut, break, or fire polish glass only by approved procedures. When inserting a glass rod or tube through a rubber or cork stopper, lubricate the glass and the stopper,

protect your hands with a portion of a lab coat or a towel, and use a gentle twisting motion.

10. **NO REMOVAL OF CHEMICALS OR EQUIPMENT FROM THE LABORATORY.** The removal of chemicals and/or equipment from the laboratory is strictly prohibited and is grounds for severe disciplinary action.

11. **NO HORSEPLAY.** Horseplay and pranks do not have a place in instructional chemistry laboratories.

12. **NO BICYCLES, ROLLER-BLADES, ETC.** Bicycles are **not allowed** in the buildings where chemistry labs meet. Using Skate Boards, In-line Skates, Roller-skates, and Unicycles is also not allowed. If skates, etc., are brought inside the building, they may not be stored by laying them on the floor.

13. **NEVER PIPET BY MOUTH.** Always use a mechanical suction device for filling pipets. Reagents may be more caustic or toxic than you expect.

EMERGENCY PROCEDURES

1. **KNOW THE LOCATION AND USE OF EMERGENCY EQUIPMENT.** Find out where the safety showers, eyewash spray, and fire extinguishers are located. If you are not familiar with the use of emergency equipment, ask your instructor for a lesson.

2. **DON'T UNDER-REACT.** Any contact of a chemical with any part of your body may be hazardous. Particularly vulnerable are your eyes and the skin around them. In case of contact with a chemical reagent, wash the affected area *immediately* and *thoroughly* with water and notify your instructor. In case of a splatter of chemical over a large area of your body, don't hesitate to use the safety shower. Don't hesitate to call for help in an emergency.

3. **DON'T OVER-REACT.** In the event of a fire, don't panic. Small, contained fires are usually best smothered with a pad or damp towel. If you are involved in a fire or serious accident, don't panic. Remove yourself from the danger zone. Alert others of the danger. Ask for help immediately and keep calm. Quick and thorough dousing under the safety shower often can minimize the damage. Be prepared to help, calmly and efficiently, someone else involved in an accident, but don't get in the way of your instructor when he or she is answering an emergency call.

These precautions and procedures are not all you should know and practice in the area of laboratory safety. The best insurance against accidents in the laboratory is thorough familiarity and understanding of what you're doing. Read experimental procedures before coming to the laboratory, take special note of potential hazards and pay particular attention to advice about safety.

Take the time to find out all the safety regulations for your particular course and follow them meticulously. Remember that unsafe laboratory practices endanger you and your neighbors. If you have any questions regarding safety or emergency procedures, discuss them with your instructor.

Investigating Mass and Volume

INTRODUCTION

Do you remember the difference between mass and weight? **Mass** is a measure of the amount of matter present and **weight** is a measure of gravitational pull. There is a similar distinction between balances and scales. The reading from a balance is a comparison of the unknown to the mass of a known amount of material. The reading from a scale is the result of the compression or expansion of a spring or similar device. Scales measure weight not mass. In this manual, weight will be commonly used to indicate mass. To determine the mass of a sample we weigh it using one of several types of balances. Your instructor will describe three common types of balances. Mass (weight) is one of the two most frequently determined quantities in the chemistry laboratory. The other commonly measured quantity is volume. In this experiment you will be given samples of two materials to use in your determination of mass and volume.

OBJECTIVES

In this experiment you will investigate the mass and volume of a material.

TECHNIQUES

Mass will be measured with precision instruments. Volume will be measured by displacement.

Correct reading of a graduated cylinder will be important to record the exact amount. When reading the volume of a solution in a graduated cylinder, you may read the units the glassware is marked off in, plus one more decimal. For example, if the glassware is marked off in milliliters, you may read to tenths of a milliliter. We are allowed one digit of estimation in significant figures. A volume of a solution is read at the bottom of the meniscus (the curved surface of the solution). A solution for which the bottom of the meniscus is exactly at the 5 mL mark should be recorded as having a volume of 5.0 mL. If the meniscus is just above the 5 mL mark, then you must estimate the last digit. You might record 5.1 mL or 5.2 mL, whichever seems the best estimation for the meniscus as it lies between the 5 mL and 6 mL marks.

EXPLORATION PROCEDURES

1. Each group of two will obtain a container in which there are two samples of material A *or* two samples of material B.

CAUTION

Handle the samples with a paper towel or tweezers to prevent putting oil from your skin on the samples!

2. Your instructor will direct you to form a larger team of three to six pairs. As this larger team, determine whether the triple beam balance, the top loader balance, or the analytical balance can yield the most significant figures for the mass of the lightest sample. Use that type of balance for the rest of this experiment. NOTE: At least 3 significant figures are needed for the mass, but depending on the sizes of your samples, your instructor may direct you to use a specific balance. When possible, the least expensive equipment should be used.

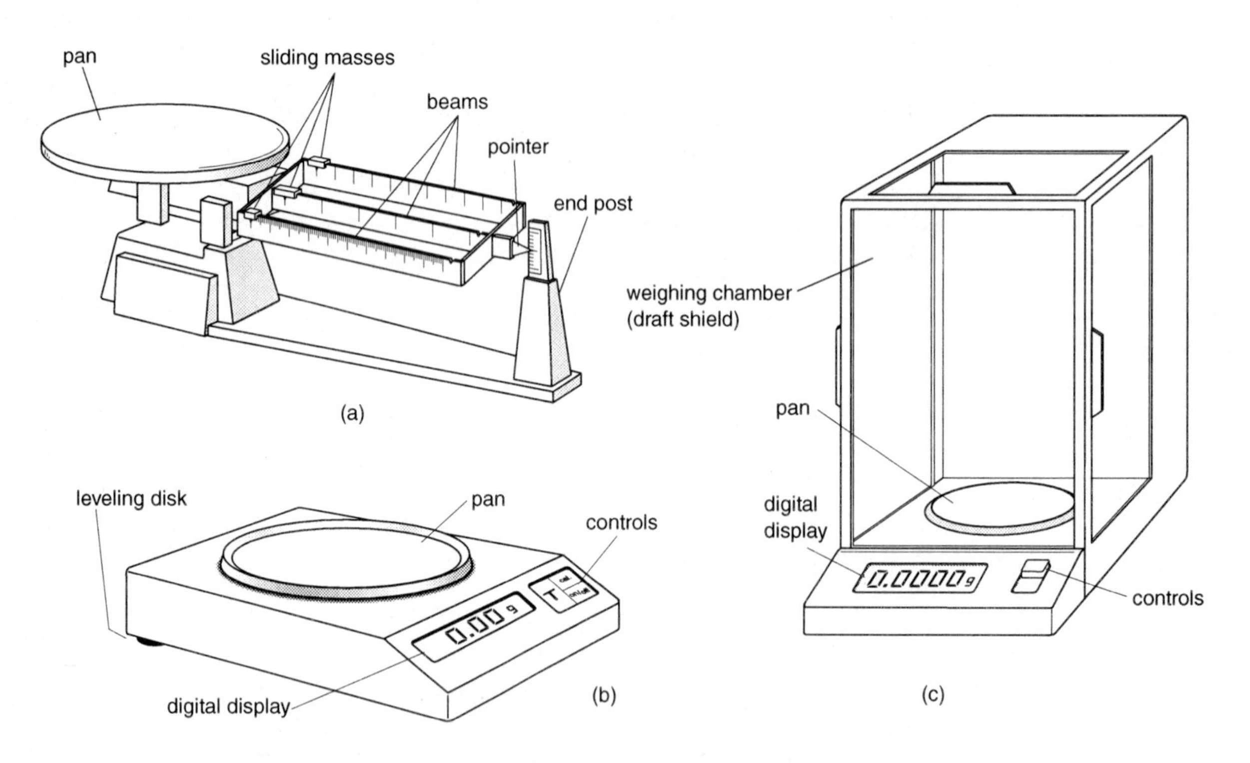

Figure 1.1 *(a) Triple beam balance* *(b) Top loader balance* *(c) Analytical balance*

3. Divide the duties within each pair so that every member weighs a sample and measures the volume of a sample. The pair must obtain two mass values and two sets of volume readings for each sample (determination #1 and determination #2).

4. Measure the volume by obtaining a 50-mL graduated cylinder. (See Figure 1.2.) Fill the cylinder with 25.0 mL of water. Using a dropper,

adjust the amount of water so that the meniscus reads exactly 25.0 mL. Tilt the cylinder and slowly slide in a sample (this will help prevent you from breaking the bottom and will prevent splashing the water). Read the new volume.

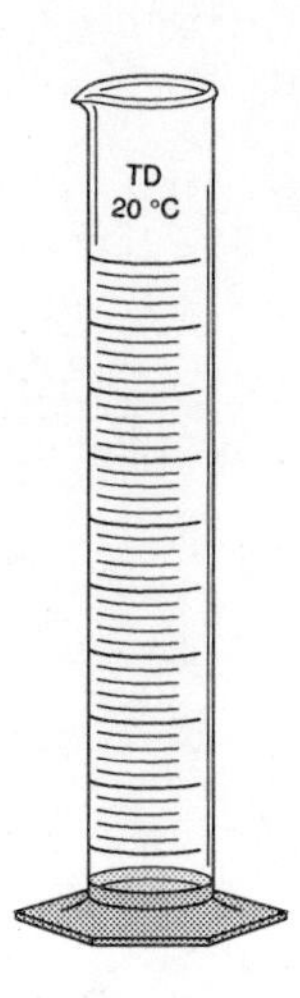

Figure 1.2 *Graduated cylinder*

5. When your group of two has obtained two mass values and two sets of dimensions for each piece, trade containers with a nearby group of two and repeat procedure #3 with this container. You should have two sets of measurements for two samples of material A and two samples of material B.
6. Next share data with other groups, so that you obtain two sets of measurements for 4 additional samples of A and 4 additional samples of B.
7. Put away all materials and clean your work area.
8. Complete the Report Form before you leave the laboratory.

Name: ______________________ Instructor: ______________________

Date (of Lab Meeting): ______________ Course/Section: ______________

Experiment 1 Prelab Exercises
Investigating Mass and Volume

1. What is meant by the term 'mass'?

2. What is meant by the term 'volume'?

3. Convert 0.2242 in. to cm. (How many digits are allowed in your answer?)

4. Why should you refrain from touching the samples used in this experiment?

Date __________ Student's Signature ______________________

Name: ______________________________

Instructor: ______________________________ Course/Section: ______________

Partner's Name: ______________________________ Date: ______________

Experiment 1 Laboratory Report Form

Investigating Mass and Volume

Data

Material A:

	Sample Number	*Mass*	*Initial Volume*	*Final Volume*	*Volume of sample (show calculation below)*
Units	-------				
Determination #1	# ____				
	# ____				
	# ____				
	# ____				
	# ____				
	# ____				
Determination #2	# ____				
	# ____				
	# ____				
	# ____				
	# ____				
	# ____				

Sample Calculations (Material A for Determination #1)

<u>*Material B:*</u>

	Sample Number	*Mass*	*Initial Volume*	*Final Volume*	*Volume of sample (show calculation below)*
Units	-------				
Determination #1	# ____				
	# ____				
	# ____				
	# ____				
	# ____				
	# ____				
Determination #2	# ____				
	# ____				
	# ____				
	# ____				
	# ____				
	# ____				

Sample Calculations (Material B for Determination #1)

(continued on back)

Date ____________ Instructor's Signature ____________________________

Analysis

1. Compare the values for material A. Obtain an average mass and an average volume for each sample of material A.

Material A:

Sample Number	*Mass (1st value)*	*Mass (2nd value)*	*Average Mass*	*Volume (1st value)*	*Volume (2nd value)*	*Average Volume*
Units						
# ____						
# ____						
# ____						
# ____						
# ____						
# ____						

2. Look for any possible relationships between mass and volume for material A. You can try using different math operations (+, –, x, /) and can try graphing. Explain what you did and what you find. Attach your graph.

3. Compare the values for material B. Obtain an average mass and an average volume for each sample of material B.

Material B:

Sample Number	*Mass (1st value)*	*Mass (2nd value)*	*Average Mass*	*Volume (1st value)*	*Volume (2nd value)*	*Average Volume*
Units						
# ____						
# ____						
# ____						
# ____						
# ____						
# ____						

4. Look for any possible relationships between mass and volume for material B. You can try using different math operations (+, -, x, /) and can try graphing. Explain what you did and what you find. Attach your graph.

Concept Questions

1. Explain the relationship between mass and volume that you found.

2. Was this relationship the same for both materials? Explain how it was the same.

3. Was this relationship different for the materials? Explain how the relationship for material A differed from that for material B.

4. What is the scientific name for the relationship you have found? If needed, discuss this with your instructor or check in your textbook.

5. Draw a view of the particles in a piece of material A, then draw a view of the particles in the same sized piece of material B.

Date ____________ Student's Signature ____________________________

Using Density to Identify an Unknown Metal Alloy

INTRODUCTION

In the previous experiment, you investigated the relationship between mass and volume. In this experiment you will be given a number of samples of pure or mixed metals. One of the samples will be an unknown. Graphical treatment of their densities may reveal a trend that can be used to identify the unknown.

OBJECTIVES

In this experiment you will conduct investigations that yield precise volume values and precise mass values. Trends in the volume, mass, or ratio of these values will (hopefully) lead to formulation of a method for analysis of an unknown sample.

TECHNIQUES

Mass will be measured with precision instruments. Volume will be measured by displacement.

APPLICATION PROCEDURES

1. Form groups of four individuals as directed by your instructor.
2. Each group is to obtain the following samples:

a. pure copper

b. pure zinc

c. three different alloys of copper and zinc

d. an unknown alloy of copper and zinc

CAUTION

Handle the samples with a paper towel or tweezers to prevent your putting oil from your skin on the samples!

3. You will measure the mass and volume of each sample. Divide the duties within each group so that every member weighs a sample and measures the volume of a sample.

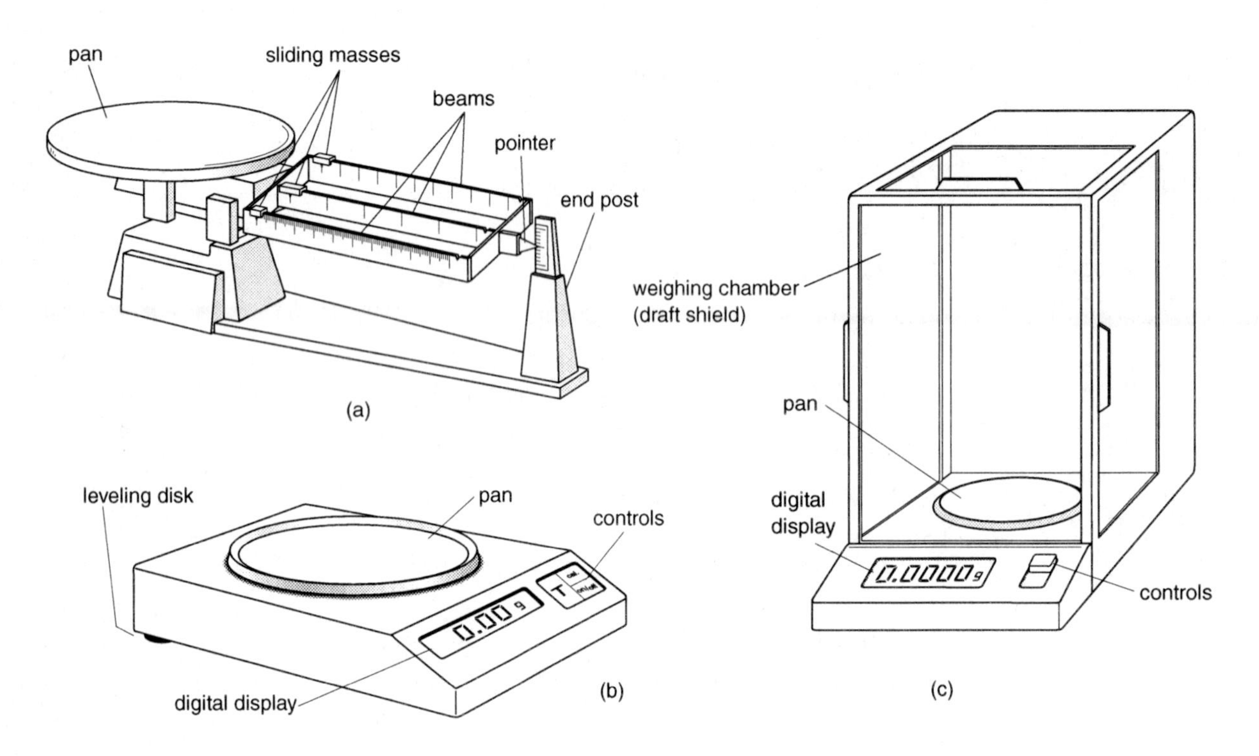

Figure 1A.1 *(a) Triple beam balance* *(b) Top loader balance* *(c) Analytical balance*

4. Measure the mass on a balance that can yield three significant figures for the mass of the lightest sample

5. Measure the volume by obtaining a 50-mL graduated cylinder. (See Figure 1A.2). Fill the cylinder with 25.0 mL of water. Using a dropper, adjust the amount of water so that the meniscus reads exactly 25.0 mL. Tilt the cylinder and slowly slide in a sample (this will help prevent you from breaking the bottom and will prevent splashing the water). Read the new volume.

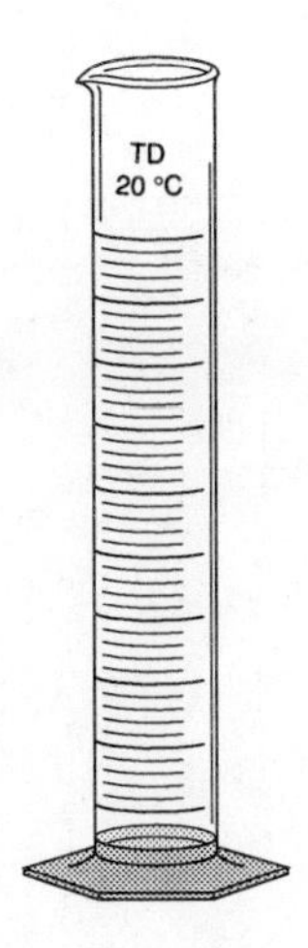

Figure 1A.2 *Graduated cylinder*

6. Each group must obtain two mass values and two sets of volume reading for each of their assigned samples including the unknown.
7. Put away all materials and clean your work area.
9. Complete the Report Form before you leave the laboratory.

Name: ______________________ Instructor: ______________________

Date (of Lab Meeting): ______________________ Course/Section: ______________________

Experiment 1A Prelab Exercises

Using Density to Identify an Unknown Metal Alloy

1. What is an alloy?

2. What is meant by the term "density"?

3. Convert 0.4435 cm to in. (How many digits are allowed in your answer?)

4. How many significant digits will be used in this experiment? Explain your answer.

Date __________ Student's Signature ______________________

Name: ______________________________

Instructor: ______________________________ Course/Section: ______________

Partner's Name: ______________________________ Date: ______________

Experiment 1A Report Form

Using Density to Identify an Unknown Metal Alloy

Data

Determination #1:

Sample Number	*Mass*	*Initial Volume*	*Final Volume*	*Volume of sample (show calculation below)*	*Density (show calculation below)*
Units					
# ____ *(% Cu)* =____					
# ____ *(% Cu)* =____					
# ____ *(% Cu)* =____					
# ____ *(% Cu)* =____					
# ____ *(% Cu)* =____					
# ____ *(% Cu)* =____					

Sample Calculations (Determination #1)

(continued on back)

<u>*Determination #2:*</u>

Sample Number	*Mass*	*Initial Volume*	*Final Volume*	*Volume of sample (show calculation below)*	*Density (show calculation below)*
Units					
# ____ (% Cu) =____					
# ____ (% Cu) =____					
# ____ (% Cu) =____					
# ____ (% Cu) =____					
# ____ (% Cu) =____					
# ____ (% Cu) =____					

Sample Calculations (Determination #2)

(continued on back)

Densities:

Sample Number	*Mass (1st value)*	*Volume (1st value)*	*Density (1st value)*	*Mass (2nd value)*	*Volume (2nd value)*	*Density (2nd value)*	*Density (average)*
# ____ (% Cu) =____							
# ____ (% Cu) =____							
# ____ (% Cu) =____							
# ____ (% Cu) =____							
# ____ (% Cu) =____							
Unknown # ____							

Analysis

1. Compare the densities to the percent copper present. Is there any observable trend? Try plotting the density versus the percent copper. Attach the graph to this report.

2. Is it possible to use the density of the unknown to identify its composition?

3. What is the likely composition of your unknown?

Concept Questions

1. Would you have obtained the same answer for the composition of the unknown if pure zinc was not included as one of the samples?

2. How do you explain that most of the data points for density are not exactly on the line that you drew for the density trend?

3. Answer other questions assigned by your instructor.

Date __________ Student's Signature ______________________

Name: ______________________________

Instructor: ______________________________ Course/Section: ______________

Partner's Name: ______________________________ Date: ______________

Experiment 1A Demo Report

Using Density to Identify an Unknown Metal Alloy

Demo #1

1. Describe the demonstration.

2. Why do some things float on some solutions but sink in other solutions?

Demo #2

1. Describe the demonstration.

2. If you have the same volume of two materials, which will be heavier, the one with a higher density or the one with a lower density. Explain your answer.

Other Demos (if any)

1. Describe what you observed during these demonstrations.

Date ___________ Student's Signature ___________________________

Investigating Properties of Materials

INTRODUCTION

Observations are the data that can be directly collected by your senses, while inferences are the conclusions that you can draw from data. You will use both observations and inferences to investigate the nature of materials in this experiment. There is a relationship between the structure of materials and their properties. Materials are selected for use by their properties. For example, a sweetener must be soluble in water and might not be marketable if its color was black.

OBJECTIVES

In this experiment you will investigate the nature of materials by observing the properties of that material.

TECHNIQUES

Critical observation and inference will be used.

EXPLORATION PROCEDURES

Individual Materials

1. Obtain two sealed vials, one of iron and one of sulfur OR one of lead and one of iodine.
2. Describe the appearance of the material in each vial.
3. Draw a magnet down the side of each vial. Record your observations about the magnetic properties of each of the materials.
4. Obtain an empty vial and record the mass of the empty vial and lid.
5. Record the mass of each sealed vial.
6. Find a group with which you can trade vials and repeat procedures 1-5 so that you have observed iron, sulfur, zinc and iodine.
7. Assuming that all empty vials have the same mass and that each of the four sealed vials are filled to the same level. Record information about the density of the four materials.

Combinations of Materials

8. Obtain two sealed vials, combination #1 and combination #2 of iron and sulfur OR combination #1 and combination #2 of lead and iodine.
9. Describe the appearance of each vial.
10. Draw a magnet down the side of each vial. Record your observations about the magnetic properties of each of the materials.
11. Find a group with which you can trade vials and repeat procedures 8-10 so that you have observed combination #1 and combination #2 of iron and sulfur AND combination #1 and combination #2 of lead and iodine.

Solubility In Hexane

12. Obtain four sealed vials. You may begin with either Set A or Set B. Record your observations for each vial in the first set.

Set A	*Set B*
hexane and iron, hexane and sulfur, hexane and combination #1 of iron and sulfur, and hexane and combination #2 of iron and sulfur	hexane and lead, hexane and iodine, hexane and combination #1 of lead and iodine, and hexane and combination #2 of lead and iodine

13. Now obtain the second set and record your observations for each vial in this second set.
14. Put away all materials and clean your work area.
15. Complete the Report Form before you leave the laboratory.

Name: ______________________ Instructor: ______________________

Date (of Lab Meeting): ______________ Course/Section: ______________

Prelab Exercises

Experiment 2 Investigating Properties

1. Give an example of an observation.

2. Give an example of an inference.

3. What is an element?

4. What is a compound?

Date ________ Student's Signature ______________________

Name: ______________________________

Instructor: ______________________________ Course/Section: ______________

Partner's Name: ______________________________ Date: ______________

Report Form

Experiment 2 Investigating Properties

Data

	Appearance	*Magnetic Properties*	*Density*	*Solubility in hexane*
Iron				
Sulfur				
Lead				
Iodine				

	Appearance	*Magnetic Properties*	*Solubility in hexane*
Iron and Sulfur #1			
Iron and Sulfur #2			
Lead and Iodine #1			
Lead and Iodine #2			

Date ____________ Instructor's Signature ______________________________

Analysis

1. Compare the data for the individual materials. How are the properties for these four materials similar, and how are they different?

2. What type of materials are iron, sulfur, lead, and iodine? Compounds or elements? Explain your reasoning.

3. Compare the data for combinations. How are the properties for iron and sulfur combination #1 and combination #2 similar and different?

4. Compare the data for combinations. How are the properties for lead and iodine combination #1 and combination #2 similar and different?

5. For iron and sulfur, which combination (#1 or #2) was most like the individual materials? Explain your reasoning.

6. For lead and iodine, which combination (#1 or #2) was most like the individual materials? Explain your reasoning.

Concept Questions

1. Which properties were made by observation, and which were made by inference?

2. If combinations #1 were mixtures, what can you say about the properties of the mixtures compared to the properties of the elements? Explain your reasoning.

3. If combinations #2 were compounds, what can you say about the properties of the compounds compared to the properties of the elements? Explain your reasoning.

4. Answer other questions assigned by your instructor.

Application Questions

1. Is air a mixture or a compound? Explain your reasoning.

2. Nitrogen dioxide (NO_2) is in a mixture with other materials to make smog. Should you be able to separate NO_2 from smog? Explain.

3. Would you expect NO_2 to have the same properties as nitrogen (N_2) and oxygen (O_2) or different properties? Explain your rasoning.

4. Would you expect the properties of oxygen gas (O_2) to be the same as ozone (O_3)? Explain your reasoning.

5. Answer other questions assigned by your instructor.

Date ______________ Student's Signature ____

Name: ______________________________

Instructor: ______________________________ Course/Section: ______________

Partner's Name: ______________________________ Date: ______________

Demo Report

Experiment 2 Investigating Properties

Demo #1

1. Describe the demonstration.

2. Describe some properties of oxygen. Are these chemical or physical properties?

Demo #2

1. Describe the demonstration.

2. Explain the differences you saw in the balloons in terms of the properties of hydrogen and oxygen.

Other Demos (if any)

1. Describe what you observed during these demonstrations.

2. Answer other questions assigned by your instructor.

Date ____________ Student's Signature ____

Trends

INTRODUCTION

The periodic table of elements contains all the known elements, both natural and synthetic. There have been a number of proposals for different arrangements of the periodic table. You will investigate the possible reactions of a number of elements with water and with a dilute solution of hydrochloric acid. Giving off bubbles or a gas is one observation that gives evidence of a chemical reaction. Next you will propose trends in the reactivity based on the element's location in the periodic table.

OBJECTIVES

In this experiment you will investigate the reactivity of elements with water and acid, then look for trends in their location in the periodic table.

TECHNIQUES

Observation, collecting a gas above water, and testing for the identity of a gas are used.

EXPLORATION PROCEDURES

1. After observing the demonstrations by your instructor, record the data from the demonstration.
2. Obtain a clean 400-mL beaker and an 18 x 150 mm test tube. Half fill the beaker with distilled water. Fill the test tube with distilled water. Holding your fingers over the mouth of the test tube, invert it into the beaker. Do not allow any air bubbles into the test tube.
3. Obtain a piece of calcium, Ca. Record observations of the appearance of calcium. Place this into the beaker. Collect any gas given off.
4. Test the liquid in the beaker with litmus paper.
5. Light a match. Carefully lift the test tube out of the water and swiftly turn the test tube horizontally while holding the match at the mouth of the test tube. Record your observations

6. Obtain three clean 18 × 150 mm test tubes.

7. Obtain pieces of magnesium, Mg, and aluminum, Al. Also obtain enough sulfur to cover the end of a spatula tip. Make observations of the appearance of these elements.

8. Place these elements in separate test tubes. Fill the tubes half way with distilled water and investigate the interactions. Record your initial observations. Then put the test tubes into a test tube rack or beaker to observe again later.

9. Obtain three additional clean 18 × 150 mm test tubes

10. Obtain pieces of magnesium, Mg, and aluminum, Al. Also obtain enough sulfur to cover the end of a spatula tip. Place these elements in separate test tubes. Fill the tubes half way with 0.1 M HCl and investigate the interactions. Record your initial observations. Then put the test tubes into a test tube rack or beaker to observe again later.

11. Record final observations of the test tubes containing water and those containing HCl.

12. Empty the solutions in the waste beaker and clean all glassware.

Name: ______________________ Instructor: ______________________

Date (of Lab Meeting): ______________________ Course/Section: ______________________

Prelab Exercises

Experiment 3 **Trends**

1. What is the periodic table?

2. What is an element?

3. Why is it necessary to have clean glassware before beginning an experiment?

4. List 3 elements you use and give their use.

Date ____________ Student's Signature ______________________________

Name: ______________________________

Instructor: ______________________________ Course/Section: ______________

Partner's Name: ______________________________ Date: ______________

Report Form

Experiment 3 **Trends**

Data

Element	*Observations of the Element and Its Reaction with Distilled Water*
Li	
Na	
K	
Ca	
Mg	
Al	
S	

Element	*Observations in Dilute Acid*
Mg	
Al	
S	

Date ____________ Instructor's Signature ______________________________

Analysis

1. Locate the elements that you investigated on the periodic table.

2. Identify any trends as you go down a column (a family) in the table.

3. Describe any patterns you see in reactivity as you go across a row (a period) in the table.

Concept Questions

1. Is there a correlation between reactivity of elements and their position in the periodic table? Explain.

2. Predict what would occur if the following were placed in water. Compare their reactivity to the elements that you observed.

   Beryllium (Be)

   Strontium (Sr)

   Chlorine (Cl)

   Rubidium (Rb)

3. What you predict about the reactivity of sodium in hydrochloric acid?

Application Questions

1. Assume that you are on a new planet, Planet Kyle. You are delighted to discover that Planet Kyle also contains water, so the elements hydrogen (H) and oxygen (O) exist. However, all other elements are new.

   Based on the following information and assuming the laws of nature are the same as those on Earth, construct a periodic table for Planet Kyle. Remember that you can *only consider relationships in columns or in rows of your periodic table.* Fill in the table below AND attach an explanation for your reasoning concerning how you placed the elements.

   Experimental Data:

	In Water
Element 1	Reacts less than 7, 2, or 3
Element 2	Reacts more than 3, but less than element 8
Element 3	Reacts more than 1, but less than 2 or 6
Element 4	Reacts more than element 6 and 2, but less than 8
Element 5	Reacts more than 3, 6, but less than 8
Element 6	Reacts less than 4 or 5
Element 7	Reacts more than 1, but less than 4 or 6
Element 8	Reacts more than all others

   Periodic Table for Planet Kyle

H			
			O

2. Which element on Kyle would you expect to be the most reactive in hydrochloric acid? Explain.

Date ____________ Student's Signature ____

Name: ____

Instructor: ____________________________________ Course/Section: ________________

Partner's Name: ________________________________ Date: ________________________

Demo Report

Experiment 3 **Trends**

Demo #1

1. Describe the demonstration and record the observations.

Demo #2

1. Describe the demonstration and record the observations.

Demo #3

1. Describe the demonstration and record the observations.

Date ____________ Student's Signature ________________________________

Conductivity

INTRODUCTION

Electricity is the flow of charges particles. Conductivity is a measure of a solution's ability to conduct electricity. You will be testing the conductivity of several solutions. Each solution will have the same number of molecular representations dissolved in the same volume of solution. Why do some conduct better than others?

OBJECTIVES

In this experiment you will investigate the conductivity of a number of dissolved compounds and reach a conclusion as to why some solutions conduct better than others.

TECHNIQUES

Observation, use of a conductivity apparatus, measuring volume, and making dilutions are used.

EXPLORATION PROCEDURES

1. Obtain two 50-mL beakers. Wash thoroughly, then rinse with distilled water and dry.
2. Obtain a stock bottle of solution labeled 0.1M $CuCl_2$. Measure 2.0 mL of the stock solution with a 10-mL graduated cylinder.
3. Measure 18.0 mL of distilled water in a 50-mL graduated cylinder (a 100-mL graduated cylinder can also be used).
4. Add both the water and the stock solution to one of the 50-mL beakers. You have now made a dilution. The new solution is 0.01M $CuCl_2$.
5. Standardize the conductivity apparatus using the instructions given to you by your instructor. Use the second beaker for the standardizing solution.
6. Rinse the probe of the conductivity meter with distilled water and gently blot the excess water. Now test and record the conductivity of your

0.01M $CuCl_2$ solution. Collect all waste and rinses in your 600-mL beaker.

7. Clean, rinse, and dry the 50-mL beakers.

8. Your instructor will designate various members of the class to obtain certain test solutions. If you are assigned to obtain a particular solution, obtain about 20 mL in a 50-mL beaker. Check the label on the stock solution. If it is not 0.01M, dilute it as needed. Place the sample on a sheet of paper with the identity and the concentration of the solution clearly written on the paper. Place your 600-mL waste beaker nearby.

9. You and your partner are to test the conductivity of each of the 12 different 0.01M solutions listed on the report form using your conductivity meter. You will need to rotate to various laboratory stations until you have tested all the solutions. Remember to rinse the meter probes after each solution.

10. When you are back at your station, empty the solutions in the waste beaker and clean all glassware.

Name: ______________________________ Instructor: ______________________________

Date (of Lab Meeting): ____________________ Course/Section: ______________________

Prelab Exercises

Experiment 4 Conductivity

1. Define an atom.

2. Define an ion.

3. What is a polyatomic ion?

4. Do all compounds contain ions? Explain your reasoning.

Date ____________ Student's Signature ______________________________

Name: ______________________________

Instructor: ______________________________ Course/Section: ______________

Partner's Name: ______________________________ Date: ______________

Report Form

Experiment 4 Conductivity

Data

Solution	*Conductivity*
KCl	
NaCl	
CH_3OH	
$MgSO_4$	
$La(NO_3)_3$	
Deionized H_2O	
$CuCl_2$	
$C_6H_{12}O_6$	
Tap H_2O	
H_2O_2	
$Al(NO_3)_3$	
C_3H_8O	
$Mg(NO_3)_2$	

Date __________ Instructor's Signature ______________________________

Analysis

1. Which solutions conducted electricity?

2. Which solutions did not conducted electricity?

3. Locate on the periodic table, each element that made up a compound, whose solution conducted electricity. Describe any patterns you see.

4. Locate on the periodic table, each element that made up a compound, whose solution DID NOT conducted electricity. Describe any patterns you see.

5. For each electrolyte or solution that DID conduct electricity, describe any patterns you see in the numerical readings.

Concept Questions

1. Why do some solutions conduct electricity and others do not?

2. Why do solutions that conduct electricity do so in different amounts? Describe how these solutions may differ and what factors may affect conductivity.

3. Can you explain the similarities and differences in conductivity in terms of dissolved species?

4. How does ionic and covalent bonding differ?

5. Choose one of the solutions and draw a view of the particles it contains.

Application Questions

1. Predict the conductivity reading you would get from the following 0.01M solutions:

   $NaNO_3$ Na_3PO_4 $AlCl_3$ CH_3CHO $Ca_3(PO_4)_2$

2. How did you make these predictions? Explain your reasoning.

3. Answer other questions assigned by your instructor

Date ____________ Student's Signature ______________________________

Name: ______________________________

Instructor: ______________________________ Course/Section: ______________

Partner's Name: ______________________________ Date: ______________

Demo Report

Experiment 4 **Conductivity**

Demo #1

1. Describe the demonstration.

2. Explain how electricity might be conducted from one person to another.

Demo #2

1. Describe the demonstration.

2. Explain the analogies presented in the demonstration.

Other Demos (if any)

1. Describe what you observed during these demonstrations.

2. Answer other questions assigned by your instructor.

Date ____________ Student's Signature ______________________________________

Energy Transfer

INTRODUCTION

Processes that involve energy transfer from samples to their surroundings are referred to as being exothermic. Those that absorb energy are endothermic. Exothermic processes cause the container or the space surrounding the sample to get warm as energy is passed from the sample to the surroundings. The burning of a match is an exothermic process, as is the cooling of water in a glass. This involves the transfer of heat energy from the water to the surroundings; therefore, also exothermic. One must consider the event from the viewpoint of the sample. In both of the previous examples the transfer of heat energy is from sample to the surroundings.

In this experiment you will observe the temperature of a sample as you add heat energy to it at a constant rate (heating with a Bunsen burner) or as you remove heat energy at a constant rate (either air-cooled or cooled in a chilled water bath). From your observations you will devise a relationship between heat energy content, temperature or phase changes, endothermic and exothermic processes, related vocabulary, *etc.*

OBJECTIVES

To conduct experiments in which heat energy is added or removed as a sample is observed and to make inferences based upon your observations. Your conclusions should reinforce your understanding of several terms and concepts.

TECHNIQUES

The Bunsen burner will be used to warm samples. The thermometer will be used to measure the temperature of samples. The proper use of both the Bunsen burner and the thermometer will be discussed by your instructor. You will need to make careful observations of temperature changes versus time. Major changes in appearance must be recorded as observed. You will be working with hot glassware and hot metal. Safe techniques are essential. Your instructor will discuss how to set up both hot and cold baths.

EXPLORATION PROCEDURES

1. Form groups of two as directed by your instructor.
2. One member of each group is to obtain a large test tube and to fill it to about 2/3 full with lauric acid. The other member is to assemble, if necessary, the hot water bath and burner. Eye protection is always worn in the chemistry laboratory.
3. Remove the stopper (if there is one) from the test tube containing the lauric acid. Place the test tube and lauric acid in the 400-mL beaker of the hot water bath. The beaker needs to be about half full with tap water. (See Figure 5.1.) Slowly heat the water in the hot water bath with a Bunsen burner at about 70°C. Do not allow the water to boil. Record your observations.

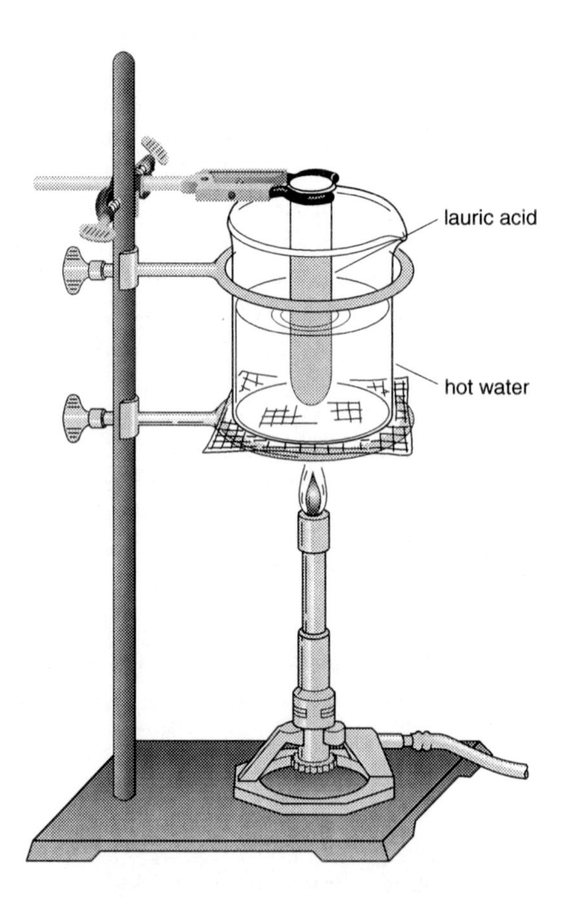

Figure 5.1 *Heating solid lauric acid*

4. One member of the team needs to prepare a beaker of cold water by placing 300 mL of tap water in another 400-mL beaker. The beaker of the water should be about 25°C. (All mercury thermometers have been replaced by non-mercury thermometers.)
5. When all the lauric acid sample has become a liquid, place a dry thermometer into the sample of melted lauric acid. When the lauric acid reaches a temperature of 65 to 70°C, remove the test tube from the hot

water bath and clamp the test tube to a ring stand. Record your observations. Proceed without pause with the next steps.

6. Hold the thermometer in the center of the liquid sample. The thermometer should not touch the bottom or the side of the test tube. Note the temperature and time. Continue noting the temperature and appearance of the sample every 30 seconds. Record your data in the table. Continue until the temperature of the sample reaches 30–35°C. The thermometer should become embedded in the sample. Then immerse the test tube with the thermometer in place into the beaker containing the 22–25°C water.

7. Reheat the hot water bath until its temperature is about 65°C. Turn off the burner. Record the temperature of the solid lauric acid and embedded thermometer. Place the test tube with solid lauric acid and embedded thermometer into the hot water bath. Do not allow the test tube to touch the bottom of the beaker. While your partner records the values, continue to observe the temperature and appearance every 30 seconds until the temperature of the sample reaches about 60°C. As soon as the solid sample breaks free of the sides of the test tube, gently mix the sample with the thermometer. If the temperature of the water falls below 60°C, heat it until the water is about 65°C.

8. Clean and return all equipment to its original place.

Name: ______________________________ Instructor: ______________________________

Date (of Lab Meeting): ____________________ Course/Section: ________________________

Prelab Exercises

Experiment 5 Energy Transfer

1. Discuss the energy flow during cooling.

2. What is meant by heat energy?

3. On the molecular level, what takes place during melting? How is energy involved with melting?

4. If liquid samples of a compound are at different temperatures, how do they differ at the molecular level?

Date __________ Student's Signature ______________________________

Name: ______________________________

Instructor: ______________________________ Course/Section: ______________

Partner's Name: ______________________________ Date: ______________

Report Form

Experiment 5 Energy Transfer

Cooling Data

Time (min)	*Temperature of sample (°C)*	*Observations*
0.0		
0.5		
1.0		
1.5		
2.0		
2.5		

Time (min)	*Temperature of sample (°C)*	*Observations*

Warming Data

Time (min)	*Temperature of sample (°C)*	*Observations*
0.0		
0.5		
1.0		
1.5		
2.0		
2.5		

Time (min)	*Temperature of sample (°C)*	*Observations*

Presentation of Data

Attach a sheet of graph paper. On the graph paper plot the temperature for the cooling data on the vertical axis and the time elapsed on the horizontal axis. Draw a smooth curve (line) through the points plotted.

On the same space on the graph paper plot the data for the warming data. Use different colors or symbols to distinguish the cooling and the warming data and curves.

Date ____________ Instructor's Signature ________________________________

Analysis

1. Where on your plots would you position the melting point? Freezing point?

2. Is melting or freezing a process that requires the removal of a quantity of energy? What on your graphs supports your answer?

3. Can you divide each of your two plots into zones? What does each zone represent and what is the shape of the curve in each zone?

4. If you assume that energy was added or removed at a constant rate during this experiment, why didn't the temperature also change at a constant rate during all zones of the warming or cooling curves?

5. What would have happened to the cooling curve of lauric acid if a larger sample had been used?

Concept Questions

1. Compare the processes referred to as melting and freezing.

2. What would you predict would be the similarities and differences between the cooling curve of lauric acid and another substance?

3. Why do different materials melt at different temperatures?

4. Answer other questions assigned by your instructor

Date ____________ Student's Signature ______________________________

Name: ______________________________

Instructor: ______________________________ Course/Section: ______________

Partner's Name: ______________________________ Date: ______________

Demo Report

Experiment 5 **Energy Transfer**

Demo #1

1. Describe the demonstration.

2. Is this process an endothermic or exothermic process? Explain your reasoning.

3. Give an example of an endothermic process that involves a phase change.

Demo #2

1. Describe the demonstration.

2. Explain the following, two samples had been setting out in the room for some time and are the same size and shape but when touched one feels colder than the other.

Other Demos (if any)

1. Describe what you observed during these demonstrations.

Date ____________ Student's Signature ______________________________________

Models

INTRODUCTION

■ **To write the Lewis Formula of a compound or polyatomic ion follow these steps:**

1. Select a reasonable (symmetrical) skeleton for the molecule or polyatomic ion.
   - **a.** The least electronegative element is usually the central atom; however, hydrogen is never the central atom.
   - **b.** The central atom is also usually the one needing the most electrons to fill its octet.
   - **c.** Oxygen atoms do not bond to other oxygen atoms except in O_3, O_2, and peroxides such as H_2O_2.
   - **d.** The skeleton is also referred to as the spider.
2. Count the valance electrons present.
3. Try to fill in the valance electrons in such a way that only single bonds and nonbonding valance electrons are present.
4. Check for compliance with the "octet rule."
   - **a.** If some atoms have less than the octet of electrons, consider the possibility of multiple bonds. Remember that most elements do not form multiple bonds.
   - **b.** If there are more electrons than needed for single bonds and nonbonding electrons to satisfy the octet rule, consider expanding the octet on the central atom.
5. Adjust the drawing to account for bonding knowledge, resonance, etc.

■ **To determine the geometry of the compound or ion follow these steps:**

1. Follow the steps above to obtain a correct Lewis Formula.
2. Count the regions of high electron density "**on**" the central atom.
3. Determine the electronic or base geometry by giving each region of high electron density its maximum space.

4. Determine the molecular (or actual) geometry by describing the shape created within the electronic geometry by only the bonded atoms.

5. Adjust the molecular geometry to account for different space needs for different sized electronic groups.

■ **To determine if a compound is polar follow these steps:**

1. Follow the steps above to obtain a correct Lewis Formula with a known geometry.

2. Ask the following 2 questions.

   a. Are there polar bonds that are not arranged so that they cancel?

   b. Are there lone pairs on the central atom that are not arranged so that they cancel?

3. If the answer to either question is yes, the compound is polar.

Table 6.1 *Table of Molecular Geometries*

Regions of electron density	*Arrangement giving maximum space (electronic geometry)*	*Number of regions used in bonding to another atom*	*Number of lone pairs*	*Molecular Geometry*
2	linear	2	0	linear
		1	1	linear
3	trigonal planar	3	0	trigonal planar
		2	1	bent
		1	2	linear
4	tetrahedral	4	0	tetrahedral
		3	1	trigonal pyramidal
		2	2	bent
		1	3	linear

OBJECTIVES

In this experiment you will practice writing Lewis Formulas and predicting their shapes

TECHNIQUES

Models will be used to visualize molecules.

EXPLORATION PROCEDURES

1. Form groups of two as directed by your instructor.
2. Your instructor will lead the class to complete a few practice molecules or ions.
3. You will be assigned one molecule or ion from each of each of the following sets.

Table 6.2 Possible Polyatomic Ions and Compounds

	List #1	List #2	List #3	List #4
Set A	1. $SiCl_4$	1. CH_3Br	1. CF_4	1. NH_4^+
Set B	2. PCl_3	2. ICl_2^+	2. ClO_3^-	2. SF_2
Set C	3. NO_3^-	3. NO_2^-	3. SO_2	3. COF_2
Set D	4. SO_4^{2-}	4. ClO_4^-	4. PO_4^{3-}	4. BrO_4^-
Set E	5. O_2	5. N_2	5. CO_2	5. CS_2
Set F	6. BF_3	6. BCl_3	6. HOOH	6. CH_3OH

4. Count the valence electrons, and write the Lewis dot structure for each.
5. Make a physical, 3-D model of each of the molecules or ions assigned. Have your instructor approve your model.
6. Determine the molecular geometry. Add the bond angles on the Lewis dot structure.
7. For each molecule, decide if your molecule is polar or nonpolar.

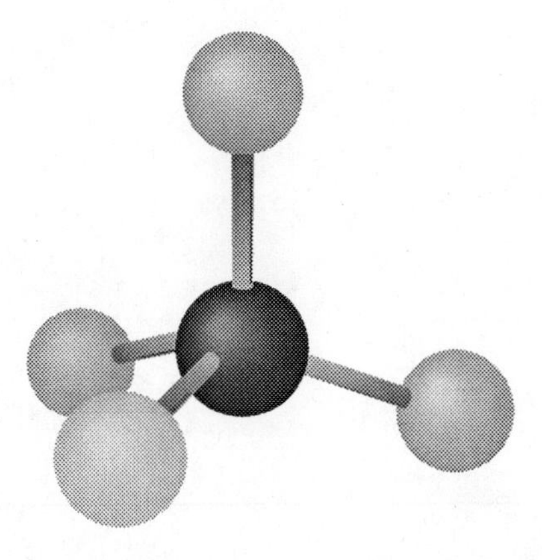

Figure 6.1 *Ball-and-stick model of methane*

Name: ______________________ Instructor: ______________________

Date (of Lab Meeting): ______________________ Course/Section: ______________________

Prelab Exercises

Experiment 6 **Models**

1. Describe what is meant by the "octet rule." You may wish to consult your textbook.

2. Describe what is meant by the "electronegativity." You may wish to consult your textbook.

3. Does a negative ion have more electrons or less electrons than the neutral atom? Explain.

4. If you have a S^{2-} ion, how many protons and electrons are present?

Date ____________ Student's Signature ______________________

Name: ______________________________

Instructor: ______________________________ Course/Section: ______________

Partner's Name: ______________________________ Date: ______________

Report Form

Experiment 6 **Models**

Give the following for each ion or molecule assigned to you:

____________ (assigned ion or molecule)	____________ (assigned ion or molecule)
1. Number of valence electrons ________	**1.** Number of valence electrons ________
2. Lewis dot structure	**2.** Lewis dot structure
3. 3-D model approved ____________	**3.** 3-D model approved ____________
4. Electronic geometry	**4.** Electronic geometry
Molecular geometry	Molecular geometry
5. Show the bond angles on Lewis structure.	**5.** Show the bond angles on Lewis structure.
6. If the assigned compound is a molecule, is it polar? Explain:	**6.** If the assigned compound is a molecule, is it polar? Explain:

Give the following for each ion or molecule assigned to you:

<table>
<tr>
<td>

_____________ (assigned ion or molecule)

1. Number of valence electrons ________

2. Lewis dot structure

3. 3-D model approved _____________

4. Electronic geometry

Molecular geometry

5. Show the bond angles on Lewis structure.

6. If the assigned compound is a molecule, is it polar? Explain:

</td>
<td>

_____________ (assigned ion or molecule)

1. Number of valence electrons ________

2. Lewis dot structure

3. 3-D model approved _____________

4. Electronic geometry

Molecular geometry

5. Show the bond angles on Lewis structure.

6. If the assigned compound is a molecule, is it polar? Explain:

</td>
</tr>
</table>

Give the following for each ion or molecule assigned to you:

____________ (assigned ion or molecule)	____________ (assigned ion or molecule)
1. Number of valence electrons ________	**1.** Number of valence electrons ________
2. Lewis dot structure	**2.** Lewis dot structure
3. 3-D model approved ____________	**3.** 3-D model approved ____________
4. Electronic geometry	**4.** Electronic geometry
Molecular geometry	Molecular geometry
5. Show the bond angles on Lewis structure.	**5.** Show the bond angles on Lewis structure.
6. If the assigned compound is a molecule, is it polar? Explain:	**6.** If the assigned compound is a molecule, is it polar? Explain:

Date ____________ Instructor's Signature ______________________________

Analysis

1. What similarities are in the molecules/ions that have tetrahedral molecular geometry?

2. What similarities are in the molecules/ions that have trigonal pyramidal molecular geometry?

3. What similarities are in the molecules/ions that have trigonal planar molecular geometry?

Concept Questions

1. Explain what factors determine the molecular geometry of a molecule or ion.

2. How can the molecular geometry be different from the electronic geometry?

3. Why was it not necessary to find the polarity of an ion?

4. Why are models used to teach some concepts?

5. Answer other questions assigned by your instructor.

Date ____________ Student's Signature ____________________________________

Name: ______________________________

Instructor: ______________________________ Course/Section: ______________

Partner's Name: ______________________________ Date: ______________

Demo Report

Experiment 6 **Models**

Demo #1

1. Describe the demonstration.

2. Draw a model of the marshmallow in the demo.

Other Demos (if any)

1. Describe what you observed during these demonstrations.

Date ____________ Student's Signature ______________________________________

Investigating Carbon Dioxide

INTRODUCTION

Carbon dioxide is one of the most familiar gases. Carbon dioxide is one of the gases that we exhale. Dry ice is solid carbon dioxide. Carbon dioxide gas becomes a solid at temperatures lower than –78.5 degrees Celsius.

OBJECTIVES

In this experiment you will investigate the properties of carbon dioxide.

TECHNIQUES

Observations and timing will be used to see how carbon dioxide behaves in a number of circumstances.

EXPLORATION PROCEDURES

1. Your instructor will demonstrate the actions of Bromothymol Blue with an acid and a base and the effect of putting a burning splint into a test tube of carbon dioxide. You will need these observations, along with the data you collect to get a good picture of the nature of carbon dioxide.
2. Obtain a 150-mL beaker half-filled with limewater. Using a clean straw blow gently into the beaker until you notice a change. Record your observations, including the time required to see a change.
3. Obtain a 150-mL beaker half-filled with distilled water. Add 4 drops of Bromothymol Blue (BTB) indicator into the water. Using a clean straw blow gently into the beaker until you notice a change. Record your observations, including the time required to see a change.
4. Obtain a small piece of dry ice in a 500-mL Erlenmeyer flask. What do you observe?
5. Place a one-hole stopper in the flask that has a short glass tube inserted in it with a rubber tube attached to the top of the glass tubing (see Figure 7.1). Place the other end of the tubing into a 150-mL beaker half-

filled with limewater. Continue until you notice a change, then remove the tubing. Record your observations.

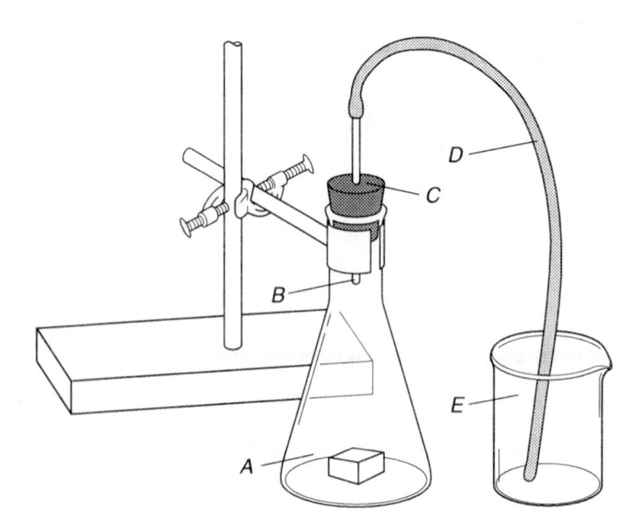

Figure 7.1 *Limewater test apparatus* ***(A)*** *500 mL Erlenmeyer* ***(B)*** *glass tubing* ***(C)*** *rubber stopper* ***(D)*** *rubber hose* ***(E)*** *150 mL beaker*

6. Fill a test tube with carbon dioxide from the end of the tubing while *holding the test tube with the mouth up*. Obtain a candle, a small piece of clay, and a match. Push the clay onto the table and insert the candle. Light the candle and quickly invert the test tube to pour the carbon dioxide onto it the burning candle. Record your results.

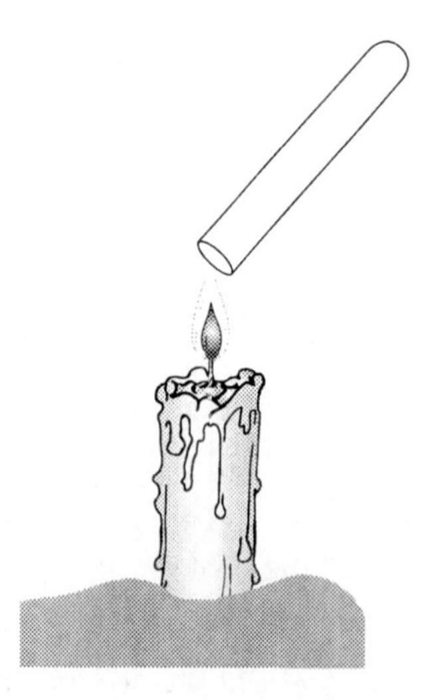

Figure 7.2 *Pouring carbon dioxide onto a burning candle*

7. Fill a test tube with carbon dioxide from the end of the tubing while *holding the test tube with the mouth down*. Light the candle and quickly pour the carbon dioxide onto it. Record your results.

8. Remove the one-hole stopper that has a rubber tube attached. Add about 100 mL of distilled water and 4 drops of Bromothymol Blue (BTB) indicator to the flask. Record your observations.

Name: ______________________ Instructor: ______________________

Date (of Lab Meeting): ______________ Course/Section: ______________

Prelab Exercises

Experiment 7 **Investigating Carbon Dioxide**

1. What is the formula of carbon dioxide?

2. What is the formula mass of carbon dioxide from the periodic table?

3. List 3 places where one might find or use carbon dioxide.

Date ____________ Student's Signature ______________________

Name: ______________________________

Instructor: ______________________________ Course/Section: ______________

Partner's Name: ______________________________ Date: ______________

Report Form

Experiment 7 Investigating Carbon Dioxide

Data

Observations of blowing into limewater, including approximate time to see a change:

Observations of blowing into water with BTB (bromothymol blue), including approximate time to see a change:

Observations of dry ice:

Observations of delivering pure carbon dioxide into limewater, including approximate time to see a change:

Observations of pouring carbon dioxide onto candles when the test tube is held with the mouth up:

Observations of pouring carbon dioxide onto candles when the test tube is held with the mouth down:

Observations of the dry ice and water with BTB, including approximate time to see a change:

Date ____________ Instructor's Signature ____

Analysis

1. What is the effect of bubbling carbon dioxide through limewater?

2. What does comparing the time required for a change in the limewater tell you about the carbon dioxide in your breath compared to the pure carbon dioxide from the dry ice?

3. What is the effect of bubbling carbon dioxide through water with BTB indicator?

4. What does comparing the time required for a change in the water with BTB indicator tell you about the carbon dioxide in your breath compared to the pure carbon dioxide from the dry ice?

5. Dry Ice is a solid. What phase changes did you observe in the dry ice placed in the flask? Did the dry ice liquefy?

Concept Questions

1. Describe the characteristic reactions of carbon dioxide with limewater and BTB water. Describe what the cloudy solution or change in color tells you.

2. What phase change does solid dry ice undergo? Is there a name for this type of change? If so what.

3. Based on your observations with the candles is carbon dioxide heavier or lighter than air? Explain.

Application Questions

1. When carbon dioxide mixes with water, what are the potential problems with solution?

2. Does all combustion produce carbon dioxide? Describe why or why not.

3. Assume that both carbon dioxide and oxygen are place into a container. Which gas would be more concentrated at the top of the container? Explain your reasoning.

Date ____________ Student's Signature ______________________________________

Name: ______________________________

Instructor: ______________________________ Course/Section: ______________

Partner's Name: ______________________________ Date: ______________

Demo Report

Experiment 7 Investigating Carbon Dioxide

Demo #1

1. Describe the demonstration.

2. What is the color of Bromothymol Blue indicator in an acid? In a base?

Demo #2

1. Describe the demonstration.

2. What are the effects of a flame inserted into carbon dioxide?

Demo #3

1. Describe the demonstration.

2. What can be concluded about the demonstration of air and of carbon dioxide?

Date ____________ Student's Signature ________________________________

From Observations to Equations

INTRODUCTION

In this and other science courses you are introduced to conclusions, concepts, theories, laws, etc., that are the result of scientists' observations. In chemistry and other molecular sciences, many of the conclusions that involve matter can be expressed via chemical equations. Chemical equations may represent a physical or a chemical change, a hypothetical or a real occurrence, or a detailed or a simplified description of the phenomena.

OBJECTIVES

In this experiment you will investigate a laboratory procedure, write an equation that represents what you believe occurred at the molecular level and perform calculations based upon your equation.

TECHNIQUES

Heating to dryness and the correct use of balances, burets, and burners are some of the techniques encountered in the experiment.

EXPLORATION PROCEDURES (Part A)

1. Form groups of two or three as directed by your instructor.
2. Obtain, clean and dry a 250-mL beaker and watch glass. Adjust an analytical balance so that it reads zero with the pan empty and the balance doors closed. Using the techniques described by your instructor, weigh the beaker plus watch glass using the analytical balance (see Figure 8.1). Record the mass to the nearest 0.1 milligram (0.0001g).

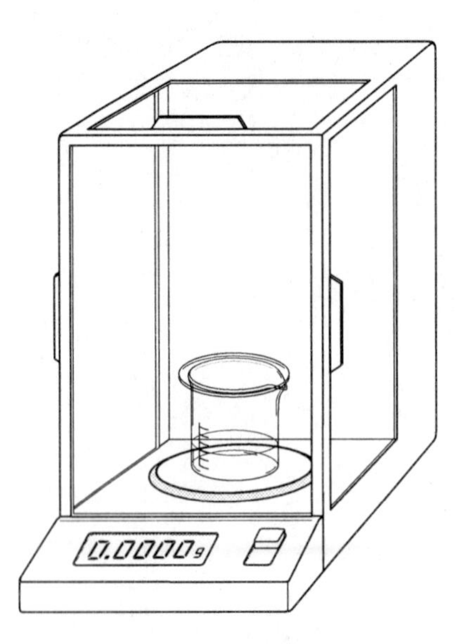

Figure 8.1 *Analytical balance with beaker plus watch glass*

3. With the beaker and watch glass still on the balance, add to the beaker between 1.0 and 3.0 grams of $NaHCO_3$ (sodium bicarbonate) by gently tapping a spatula containing sodium bicarbonate with your free hand. Record the mass of beaker and watch glass plus $NaHCO_3$ to the nearest 0.1 milligram ((0.0001g). Leave the balance and the area around it clean.

4. Obtain a clean 50-mL buret. In a small beaker obtain about 60 mL of 1 *M* hydrochloric acid (HCl) solution. Use a few milliliters of the acid solution to rinse the buret. Drain the buret and fill it with the acid solution. Drain a few milliliters of solution to insure all the air is removed from the buret tip. Record the volume in the buret (see Figure 8.2). Slowly add 10 mL of the acid solution to the beaker containing the $NaHCO_3$. Swirl the beaker and observe the results (see Figure 8.3).

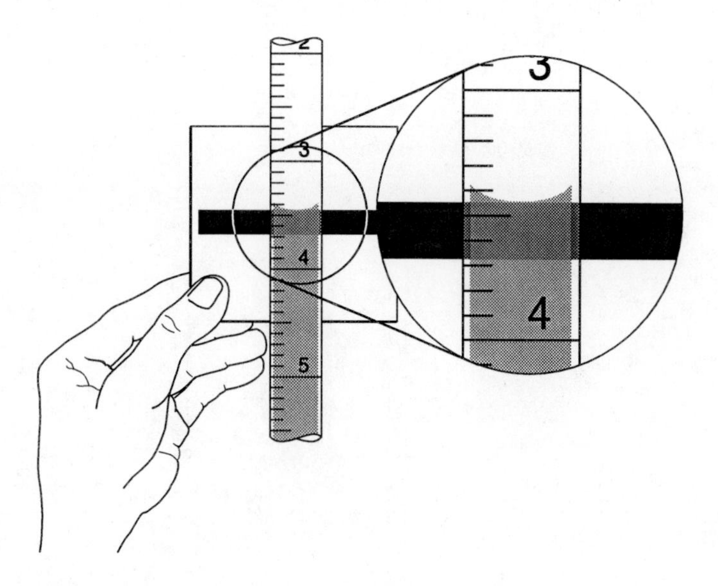

Figure 8.2 *Reading a buret (a volume of 3.44mL is show)*

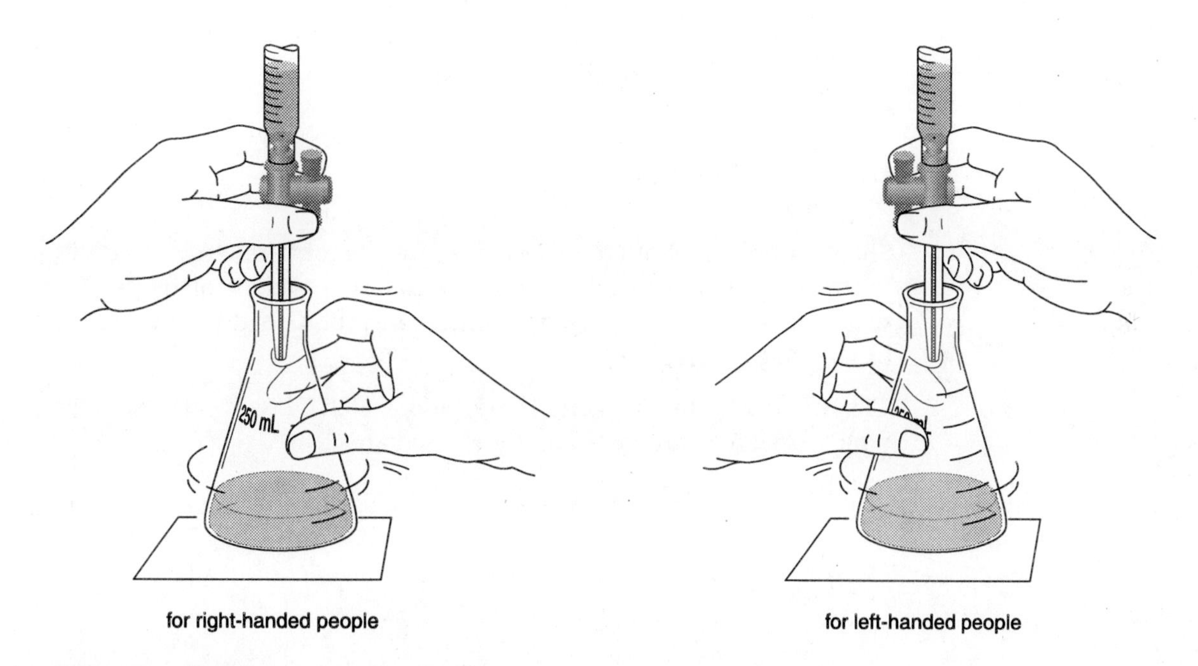

Figure 8.3 *Dispersing a liquid with a buret*

5. Continue to add the acid solution to the beaker containing the $NaHCO_3$ until no further change is observed. Add the acid very slowly at the end. Record the final volume contained in the buret.

6. Put the watch glass back on the beaker. Place the beaker and watch glass on a ring stand, secure with proper rings (see Figure 8.4). Heat the contents of the beaker to dryness. Allow the beaker to cool and weigh the beaker, watch glass, and dry residue. To check for dryness of the residue, return the beaker, watch glass, and residue to the ring stand and heat a few minutes. Allow everything to cool. Reweigh the

beaker, watch glass and residue. If there is no change in weight, the residue is considered to be dry. Record your final mass determination.

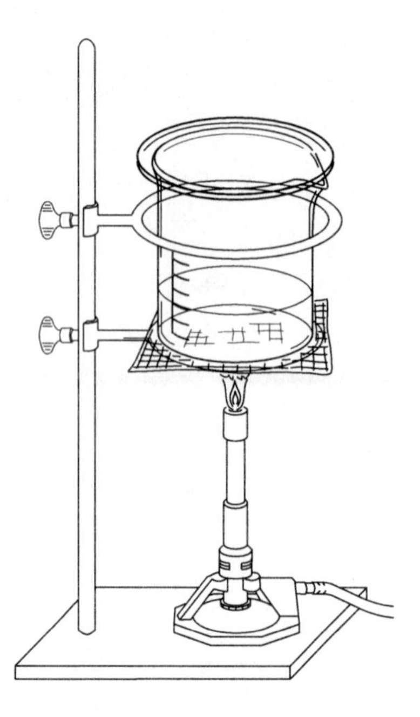

Figure 8.4 *Heating apparatus*

7. Share with the rest of your class the value you obtain when you divide the mass of residue you obtained (minus the weight of beaker and watch glass) by the mass of sodium bicarbonate used (minus the beaker and watch glass).
8. Clean and dry the beakers, buret, and watch glass. Make sure that your work area and the balance you used are clean.

EXPLORATION PROCEDURES (Part B)

1. Your instructor will have on display the 8 test tubes that contain the following:

Test tube #	$Co(NO_3)_2$	Na_3PO_4	*Total Volume*
1	0.15 grams	0.026 grams	12 mL
2	0.15 grams	0.052 grams	12 mL
3	0.15 grams	0.078 grams	12 mL
4	0.15 grams	0.104 grams	12 mL
5	0.15 grams	0.130 grams	12 mL
6	0.15 grams	0.156 grams	12 mL
7	0.15 grams	0.0 grams	12 mL
8	0.00 grams	0.156 grams	12 mL

2. Find the moles of each compound in tubes 1-8. Record these values on your report form and show sample calculations in your notebook.
3. Record your observations of the precipitate (the residue) and the supernatant liquid.

Name: ______________________ Instructor: ______________________

Date (of Lab Meeting): ______________ Course/Section: ______________

Prelab Exercises

Experiment 8 From Observations to Equations

1. Write a balanced chemical equation that describes the reaction of hydrogen (H_2) with oxygen (O_2) to yield water (H_2O).

2. In terms of chemicals changing, describe what you would expect to observe during the reaction described in question #1.

3. Describe the quantity "mole."

4. During the reaction described in question #1 is there mass balance? Explain.

Date __________ Student's Signature ______________________

Name: ______________________________

Instructor: ______________________________ Course/Section: ______________

Partner's Name: ______________________________ Date: ______________

Report Form

Experiment 8 From Observations to Equations

Data (Part A)

Mass of beaker + watch glass ____________ grams

Mass of beaker + watch glass + $NaHCO_3$ ____________ grams

Mass of beaker + watch glass + dry residue ____________ grams

Observation(s) that indicated the occurrence of a reaction.

Initial reading on buret ____________ mL

Final reading on buret ____________ mL

Volume of 1.0 *M* hydrochloric acid used ____________ mL

Class data: Mass of $NaHCO_3$, mass of residue, and ratio of mass of residue to mass of sodium bicarbonate used.

Group	*Mass of $NaHCO_3$*	*Mass of residue*	*Ratio*	*Group*	*Mass of $NaHCO_3$*	*Mass of residue*	*Ratio*

Data (Part B)

Observation(s)

Tube	*Observation of liquid*	*Observation of precipitate*
1		
2		
3		
4		
5		
6		
7		
8		

Moles of reactants:

Test tube #	*$Co(NO_3)_2$*	*Na_3PO_4*
1	______ moles	______moles
2	______ moles	______moles
3	______ moles	______moles
4	______ moles	______moles
5	______ moles	______moles
6	______ moles	______moles
7	______ moles	______moles
8	______ moles	______moles

Date ____________ Instructor's Signature ________________________________

Analysis (Part A)

1. Circle the best description below, then explain your choice.

   $NaHCO_3$ is reacting with HCl. OR $NaHCO_3$ is dissolving in HCl.

2. What are the possibilities for the gas involved in the bubbling you observed? Which do you think is the best choice? Explain your choice.

3. What are the possibilities for the formula of the residue? Which do you think is the best choice? Explain your choice.

4. Plot the mass of the residue on the y-axis and the mass of the $NaHCO_3$ on the x-axis. Consider if 0.0 should be included or forced. Give the algebraic equation for your graph. Attach the graph to your report.

5. How are the ratios of the mass of the residue to the mass of sodium bicarbonate related to your graph?

6. Write a balanced chemical equation for the possible products of the combination of HCl and $NaHCO_3$.

7. Using the periodic table, what is the ratio of the mass of the formula of your proposed residue to the molar mass of $NaHCO_3$?

8. Should the ratio of the masses from the periodic table be the same as those used in lab? Explain.

9. Based on all evidence, what is the balanced equation. Explain your reasoning.

Concept Questions (Part A)

1. What can you say about the moles of the residue compared to the moles of $NaHCO_3$?

2. Is the relationship you described in the previous question constant or varying? Explain.

3. What can you say about the mass of the residue compared to the mass of sodium bicarbonate?

Application Questions (Part B)

1. Write a proposed balanced equation for the reaction of $Co(NO_3)_2$ and Na_3PO_4.

2. Discuss the differences in the appearance of the precipitate and the supernatant liquid in tubes 1 through 8.

3. Predict the number of moles of Na_3PO_4 needed to react with all of the $Co(NO_3)_2$. How many grams of would this be?

4. In which tube was one of the reactants used completely up?

5. In a balanced equation, is there a relationship between the reactant and products in the number of moles, the mass, or the number of atoms. Explain each of the three. (Are these three the same for the reactants vs. the products?)

Date ____________ Student's Signature ______________________________________

Name: ______________________________

Instructor: ______________________________ Course/Section: ______________

Partner's Name: ______________________________ Date: ______________

Demo Report

Experiment 8 **From Observations to Equations**

1. Describe demonstration #1.

2. Describe demonstration #2.

3. Describe demonstration #3.

4. Write an equation that describes the chemical change taking place in one of the demonstrations.

5. What evidence did you observe that lead you to conclude that a chemical change was taking place in any of the demonstrations?

Date ____________ Student's Signature ____________________________________

Identification of a Substance

INTRODUCTION

In this experiment you are to determine some of the chemical properties of six unknown substances when these are dissolved in water and labeled as A, B, C, D, E, and F. A chemist would try to establish the identity of each and to illustrate each chemical change with an equation. For this experiment you are to let them remain as compounds A, B, C, D, E, and F and to describe in words (not equations) all chemical changes you observe as various pairs of these solutions are mixed. Using your observations, you are to identify two unknowns that are the same as two of the solutions.

OBJECTIVES

In this experiment you can envision a professor going to a storage cabinet where several bottles of solutions labeled as aqueous solutions of A, B, C, D, E, and F are stored. To the professor's surprise, several of the labels have come off. Using the solutions from bottles whose labels were still attached, the professor was able to observe the chemical changes, if any, that occur when the pairs indicated in Table 1 on the Report Form were mixed. In this experiment you are to conduct similar experiments and to construct a table similar to Table 1 and use tests with the known solutions to identify each of two unknowns as being the same as A, B, C, D, E, or F.

TECHNIQUES

Using a spot plate (a well plate may also be used), first try placing 10 drops of one of the known solutions in a depression on the spot plate (see Figure 8A.1). Slowly add 10 drops of a second solution. Mixing may not be necessary, so observe the mixture as each drop is added. After the 10 drops have been added, carefully mix. Record your observations in your notebook. You may need to place the spot plate on a dark surface in order to see any white precipitate formed. Repeat with a second combination of solutions. Continue to test combinations and to record your observations in your notebook until you have collected the data needed for the completion of Table 1.

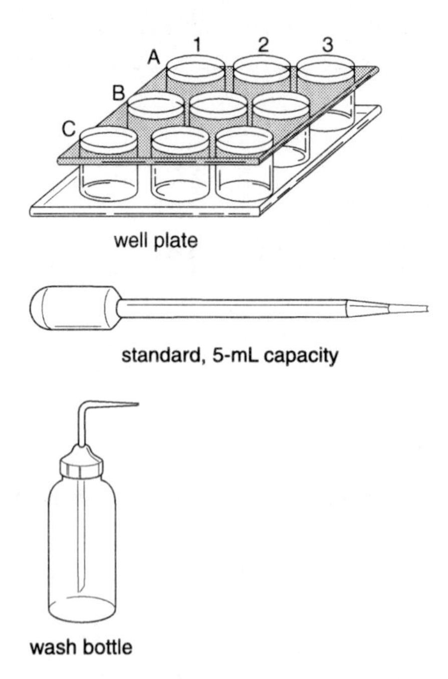

Figure 8A.1 *Use a standard beral pipet to fill the well or spot plate. Use a wash bottle to clean the plate*

Use all the spots on the spot plate before you take it and a wash bottle to the waste container. Rinse the solutions with a small amount of water from the wash bottle into the funnel on the waste container. Most of the washing and drying of the spot plate is to be done at the sinks. Only the initial rinse after each use of the spot plate is to be collected in the waste container.

APPLICATION PROCEDURES

1. Using the techniques described above, run the tests needed for the completion of Table 1. Record observations in your notebook.
2. Keep the amounts small for all tests.
3. Test your two assigned unknowns with each of the six known solutions.
4. Using the results recorded in your notebook, fill out Table 1.
5. Complete the Report Form. Your Report Form will need to include:

   Problem Statement: This includes a few sentences describing the specific question(s) you are trying to answer with your experiment.

   Procedures: This section contains the materials and equipment that you will use, the type of data you will collect (the variables you will measure), and the number of trials you are proposing. Remember to discuss safety considerations.

Data/Analysis: Include the data you collected. Data should be in tables when possible, with easy-to-read labels. Analysis of the data should also be included (analysis = what your data tells you).

Conclusion: This is the generalization or explanation you have deduced from your experiment. This is also the place to make explanations for any data results that are counter to logical chemical ideas and to describe how you would change the experiment if you repeated it.

Name: ______________________ Instructor: ______________________

Date (of Lab Meeting): ______________ Course/Section: ______________

Prelab Exercises

Experiment 8A Identification of a Substance

1. Problem Statement

2. If a solution of compound A forms a white solid when mixed with a solution of compound B, what would you expect to observe when 10 drops of each solution are mixed in a depression on a spot plate?

3. Is the formation of a gas an indicator of a chemical reaction taking place? Why or why not?

4. Should all the liquids form a solid when mixed with a second liquid? Why or why not?

5. Do solids form slowly or very rapidly? Why or why not?

6. When a gaseous product is formed by a reaction, is one large bubble or many smaller bubbles formed? (Hint: What do you observe when a carbonated beverage is poured into a glass?)

7. Define:
   a. Gaseous product

   b. Precipitate

8. Match the terms in the first column with the abbreviations in the second column.

______	a.	precipitate	1.	aq
______	b.	unknown	2.	rxn
______	c.	solution	3	ppt
______	d.	reaction	4.	soln
______	e.	aqueous	5.	unk

Date ____________ Student's Signature ______________________________

Name: ______________________________

Instructor: ______________________________ Course/Section: ______________

Partner's Name: ______________________________ Date: ______________

Report Form

Experiment 8A **Identification of a Substance**

Data

Data is collected in your notebook.

Date __________ Instructor's Signature ______________________________

Lab Report

Attach this sheet to your lab report that includes the **Problem Statement** (actual), **Procedures** (actual), **Data/Analysis**, and **Conclusions/Concept Questions**.

Problem Statement: This includes a few sentences describing the specific question(s) you are trying to answer with your experiment.

Procedures: This section contains descriptions of your experimentation, the materials and equipment that you used, the type of data you collected (the variables measured), and the number of trials you ran. Remember to discuss safety considerations.

Data/Analysis: After completing the data collection, you will write a lab report. Although you collected data and shared ideas with a partner, you are expected to write the final lab report independent of your partner. Your grade will depend on the thoroughness of your investigation, the presentation of your data, the careful analysis of the data, and the logic put in to give reasonable results and explanations.

Conclusions/Concept Questions: The lab report MUST include the tables found on the following pages. You should discuss the identities of your unknowns and state a summary of the evidence for your claim. Discuss any errors or changes you would propose to your procedure or analysis. Attach or include the Concept Questions.

Data

Table 1

(Indicate in the following table the results obtained when 10 drops of the solution shown at the right end of the row are added a drop at a time to 10 drops of the solution shown at the top of the column.)

soln A	*soln B*	*soln C*	*soln D*	*soln E*	*soln F*	
						soln A
						soln B
						soln C
						soln D
						soln E
						soln F

Your Unknowns (10 drops of each of your unknowns plus 10 drops of the designated unknown.)

	soln A	*soln B*	*soln C*	*soln D*	*soln E*	*soln F*
Your unknown #________						
Your unknown #________						

Date ____________ Instructor's Signature ____________________________________

Analysis

Table 2

Use the results shown in Table 1 to complete Table 2.
(You can make the presumption that the order in which the solutions are added will not change the observations made in this experiment.)

soln A	*soln B*	*soln C*	*soln D*	*soln E*	*soln F*	
No rxn						*soln A*
	No rxn					*soln B*
		No rxn				*soln C*
			No rxn			*soln D*
				No rxn		*soln E*
					No rxn	*soln F*

Which of the original unknowns are the same as your unknowns?

	Unknown's Identity
Your unknown #__________	
Your unknown #__________	

Concept Questions

1. Describe the reasons that each of your unknowns could not be any of the non-selected solutions.

2. One of the known solutions in this experiment may have been a solution of sodium chloride, common table salt. Sodium chloride as either a solid or an aqueous solution does not react with most other compounds. However, one reaction that it does undergo is:

$$NaCl(aq) + AgNO_3(aq) \rightarrow AgCl(solid) + NaNO_3(aq)$$

   a. Based upon this information and the information you gathered during this experiment, which solution (A, B, C, D, E, or F) was a solution of sodium chloride?

   b. Which solution was a solution of silver nitrate?

3. KI has many properties similar to NaCl. For example it forms a creamy white ppt. with $AgNO_3$. How is this property similar or dissimilar to a property of NaCl?

Date ____________ Student's Signature ______________________________________

Name: ______________________________

Instructor: ______________________________ Course/Section: ______________

Partner's Name: ______________________________ Date: ______________

Demo Report

Experiment 8A Identification of a Substance

1. Describe what you observed during demonstration #1.

2. Describe what you observed during demonstration #2.

3. Describe what you observed during demonstration #3.

Date ____________ Student's Signature ______________________________

Nature of Reactions

INTRODUCTION

In this experiment, you will be combining tincture of iodine, vitamin C, and bleach to observe the characteristics of a reaction. The tincture of iodine solution from the drug store is really a solution of I_2 molecules. Tincture of iodine was once a very popular substance to put on minor wounds (antibiotics are now favored as better "germ-killers"). The iodine solution is brownish-red in color, while the iodine ion (I^-) in solution is colorless. Vitamin C is ascorbic acid. It is sour to taste, like all acids, and is found in citrus products. Household bleach is a solution of sodium hypochlorite (NaOCl).

OBJECTIVES

In this experiment you will investigate a series of combinations to determine the reaction and the factors that affect the speed of the reaction.

TECHNIQUES

The important techniques in this experiment are the correct use of a graduated cylinder, and observation skills, including timing.

EXPLORATION PROCEDURES

1. Obtain 5 tablets of vitamin C.
2. Place 1 tablet on a paper towel. Using the dropper in the iodine container, carefully add 1 drop of a tincture of iodine solution to the tablet. Observe the drop for 2–3 minutes. Record in your notebook the time and description of any changes. You will use your notes later to fill out your report form.
3. Add 1–2 drops of bleach to the iodine spot on the tablet. Observe for 1–2 minutes. Record the time and description of any changes.
4. Place a new tablet of vitamin C on the paper towel. Write on the towel to label the 2nd tablet. Carefully, add 1–2 drops of bleach to the 2nd

tablet. Observe the drops for 2–3 minutes. Record the time and description of any changes.

5. Place the 3rd tablet in a mortar and use the pestle to grind it up. Pour the powdered tablet onto a paper towel, shaping the powder into a rectangle with a flat top. Using the dropper in the iodine container, carefully add 1 drop of a tincture of iodine solution to the powder. Record in your notebook the time and description of any changes.

6. Add 1–2 drops of bleach to the iodine spot on the powder. Observe for 1–2 minutes, recording the time and description of any changes.

7. Place the 4th vitamin C tablet on a white paper towel.

8. Obtain a small test tube; add 2 drops of iodine solution. Then add 2 drops of distilled water, using your graduated cylinder. Stir with a glass rod.

9. Use a dropper to add 1-drop of the iodine solution from the test tube to the 4th tablet. Observe the drop for 1–2 minutes. Record the time it takes for any change to occur and your observations of the change.

10. Add 1–2 drops of bleach to the spot where you placed the iodine solution. Observe for 1–2 minutes, recording the time and description of any changes. Dispose of the iodine solution that you made in the test tube as your instructor directs.

11. Obtain a clean, dry 50-mL beaker. Add 25 mL of distilled water, using a graduated cylinder. Add the 5th vitamin C tablet to the water in the beaker and stir with a glass rod. After some of the table dissolves (it is not necessary to dissolve the complete tablet), remove the remaining tablet.

12. Place the beaker on a white paper towel. Add 1 drop of the tincture of iodine to the beaker. Observe the beaker until you see the same changes as with the 1st tablet. Record your observations, including the time for any changes to occur.

13. Add 1–2 drops of bleach. Observe the beaker and record the time and description of any changes.

14. Add tincture of iodine a drop at a time until a faint yellow color persists for 20 seconds. You can repeat the addition of the 1–2 drops of bleach and/or the tincture of iodine. You are encouraged to experiment.

15. Empty, clean, and dry your glassware as your instructor directs.

Name: ______________________________ Instructor: ______________________________

Date (of Lab Meeting): ____________________ Course/Section: ____________________

Prelab Exercises

Experiment 9 Nature of Reactions

1. If you have a solution of iodine that is brownish red, what form of iodine (the molecule or the ion) would you predict to be in the solution? Explain the difference.

2. What is the purpose of placing a beaker on a white paper towel?

3. If you are using a graduated cylinder that has 1-mL increments to measure distilled water and the meniscus is just under the 16-mL mark, what volume should you record? Explain your answer.

4. What is another name for vitamin C? Give 2 household products that contain vitamin C.

5. Answer any question provided by your instructor.

Date ____________ Student's Signature ______________________________________

Name: ______________________________

Instructor: ______________________________ Course/Section: ______________

Partner's Name: ______________________________ Date: ______________

Report Form

Experiment 9 Nature of Reactions

Data

	After Addition of Iodine (time and observations of any change)	*After Addition of Bleach (time and observations of any change)*
1st Tablet		
2nd Tablet		
3rd Tablet (powdered)		

	After Addition of Iodine (time and observations of any change)	*After Addition of Bleach (time and observations of any change)*
4th Tablet	With your solution prepared in Procedure 8.	
5th Tablet in solution		

Describe your experiments with the solution from tablet 5.

Date ____________ Instructor's Signature ______________________________

Analysis

1. Did the tincture of iodine react with the vitamin C tablet? If there was a reaction, what was one of the products? Explain what observation(s) you used for your answers.

2. Compare results for the 1st and 2nd tablets. Did the bleach react with the vitamin C tablet or the products of the iodine/vitamin C combination? Explain your reasoning.

3. What form was the iodine in before AND after the application of bleach? Explain, using your observations.

4. Consider the 1st and 3rd tablets. What variable(s) was (were) changed? What can you conclude from your data on these tablets? Give the evidence to support your answer.

5. Considering the 1st and 4th tablets. Does the concentration of the iodine solution make a difference? Give the evidence for your answer.

6. Considering the 1st and 5th tablets. What variable was changed? What effect did this have on the reaction? How is this related to the 3rd tablet?

7. From the demonstration concerning temperature: Did a warmer solution make a difference? What is the effect of temperature? Give the evidence for your answer.

Concept Questions

1. Did any reaction take place that was reversed in this experiment? Explain.

2. What are some factors that affect the speed of a reaction?

3. Draw a particle view of the iodine solution before it was added to the vitamin C tablet and after it reacted.

Application Questions

1. A student finds that 3 grams of baking soda (a solid) and 15-mL of vinegar (5% acetic acid) react together by bubbling vigorously. The student wants to slow this reaction down. List at least two ways that he/she could do this.

Date ____________ Student's Signature ____

Name: ______________________________

Instructor: ______________________________ Course/Section: ______________

Partner's Name: ______________________________ Date: ______________

Demo Report

Experiment 9 Nature of Reactions

1. Describe demonstration #1.

2. Can you explain the results you observed?

3. Describe demonstration #2.

4. Can you explain the results you observed?

5. Describe any other demonstration.

Date ____________ Student's Signature ____________________________________

Nature of Substances

INTRODUCTION

The operational characteristic or definition of a substance describes how the substance operates in everyday life or in the laboratory. This macroscopic behavior is observable. For example, you can see that baking soda bubbles when vinegar is added. A theoretical characteristic or definition involves abstract, non-observable traits of a substance. For example, baking soda contains sodium, hydrogen, carbon, and oxygen. You will find the operational characteristics of a number of solutions and household materials. The pH of a substance is one characteristic. The pH is a scale from 0 to 14 that measures the hydrogen ion content of a substance. Deionized water is "pure" water with no other ions or dissolved substances.

OBJECTIVES

In this experiment you will investigate the reactions of a number of substances with magnesium, magnesium nitrate, bromothymol blue, phenolphthalein, cabbage juice, and pH paper and then use the results to look for patterns in the results.

TECHNIQUES

Correct use of pH paper and indicators will be used.

EXPLORATION PROCEDURES

1. Obtain a clean, dry spot plate or well plate. Place the plate on a piece of paper.

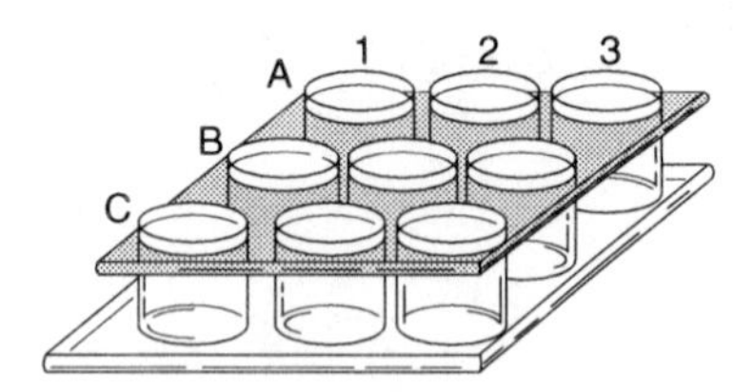

Figure 10.1 Well plate

2. Obtain four pieces of red litmus, and four pieces of blue litmus. Divide each piece into thirds, so that you now have twelve pieces of each. Obtain 12 small pieces of pH paper. Your instructor may give you a larger piece that you need to divide.
3. Fill 6 spots 3/4 full with 0.1M HCl.
4. Add a small piece of magnesium into one of the spots. Record your observations.
5. Into another spot, add 3 drops of 1*M* $Mg(NO_3)_2$. Stir with a toothpick. Record your observations.
6. Into a new spot, dip in a small piece of red litmus paper. Record your observations. Then dip a small piece of blue litmus paper into the spot. Record your observations. Finally, dip a small piece of pH paper into the spot. Compare the color obtained with the chart provided. Record your observations.
7. Into a 4th spot, add 2 drops of phenolphthalein (PHN). Record your observations.
8. Into a 5th spot, add 2 drops of bromothymol blue (BTB). Record your observations.
9. Into a 6th spot, first measure the conductivity, then add 2 drops of cabbage juice. Record your observations.
10. Clean and dry the plate. Repeat steps 3–9 with deionized (DI) water.
11. Repeat steps 3–9 with 0.1 M NaOH, HNO_3, $Ba(OH)_2$, $HC_2H_3O_2$, KOH, H_2SO_4. Be sure to clean and dry the plate between each substance.
12. Test 4 household substances by repeating steps 3–9, remembering to THOROUGHLY clean the plate between samples. If the substance is a solid, dissolve it in a small amount of deionized water.
13. Clean and dry the glassware. Make sure that your work area and the balance you used are clean.

Name: ______________________________ Instructor: ______________________________

Date (of Lab Meeting): ______________________ Course/Section: ______________________

Prelab Exercises

Experiment 10 Nature of Substances

1. From its place in the periodic table, what kind of material is magnesium?

2. What is a precipitate?

3. Indicate for which combinations there is a strong possibility that a chemical reaction has taken place. Put RXN for reaction and NR for no reaction

______ Two solutions are combined, and a precipitate is formed.

______ A solution and a solid are combined, and bubbling begins.

______ A clear, colorless solution and a bright yellow solution are combined, and the result is a pale yellow solution.

______ A white solid and a black solid are combined, and the result is a "salt and pepper" solid.

______ A pale pink solution and a clear, colorless solution are combined, and the result is a bright blue solution.

4. Define operational characteristic.

5. Define theoretical characteristic.

Date ____________ Student's Signature ______________________________

Name: ______________________________

Instructor: ______________________________ Course/Section: ______________

Partner's Name: ______________________________ Date: ______________

Report Form

Experiment 10 Nature of Substances

Data

	HCl	*H_2O (DI)*	*NaOH*	*HNO_3*
Mg				
$Mg(NO_3)_2$				
Red Litmus				
Blue Litmus				
pH Paper				
PHN				
BTB				
Cabbage Juice				
Conductivity				

	$Ba(OH)_2$	$HC_2H_3O_2$	KOH	H_2SO_4
Mg				
$Mg(NO_3)_2$				
Red Litmus				
Blue Litmus				
pH Paper				
PHN				
BTB				
Cabbage Juice				
Conductivity				

Household Materials:

	Material Tested ______	*Material Tested* ______	*Material Tested* ______	*Material Tested* ______
Mg				
$Mg(NO_3)_2$				
Red Litmus				
Blue Litmus				
pH Paper				
PHN				
BTB				
Cabbage Juice				
Conductivity				

Date ____________ Instructor's Signature ______________________________

Analysis

1. Look for patterns in the eight substances that you first tested: HCl, H_2O (DI), NaOH, HNO_3, $Ba(OH)_2$, $HC_2H_3O_2$, KOH, H_2SO_4. How many patterns do you see? Describe the patterns.

2. List the substances that tested the same as HCl.

3. What are operational characteristics or definitions for the substances that tested the same as HCl? (What test results did these substances give?)

4. The substances that tested the same as HCl are called acids. What similarity do you see in the formulas for acids? Would similarity in formula be an operational or theoretical characteristic?

5. List the substances that tested the same as NaOH.

6. What are operational characteristics or definitions for the substances that tested the same as NaOH? (What test results did these substances give?)

7. Substances that tested the same as NaOH are called bases. What similarity do you see in the formulas for bases? Would similarity in formula be an operational or theoretical characteristic?

8. List the substances that tested differently than HCl or NaOH.

9. What are operational characteristics or definitions for these substances? (What test results did these substances give?) Circle the tests that are the same as acids and box the tests that were the same as bases. These substances are called neutral substances.

Concept Questions

1. Give the operational and theoretical definitions of acids.

2. Give the operational and theoretical definitions of bases.

Application Questions

1. List each of your 4 household materials. Then indicate if they are acidic, basic, or neutral and explain your reasoning.

   Material I: ___________________

   Material II: ___________________

   Material III: ___________________

   Material IV: ___________________

Date _____________ Student's Signature ____

Name: ______________________________

Instructor: ______________________________ Course/Section: ______________

Partner's Name: ______________________________ Date: ______________

Demo Report

Experiment 10 **Nature of Substances**

1. Describe demonstration #1.

2. Explain what substance in your breath probably caused the change in the liquid solution.

3. Describe any other demonstration.

Date ____________ Student's Signature ______________________________

Reactions of Acids and Bases

INTRODUCTION

You have explored the nature of acids and bases by examining how they react with litmus paper, pH paper, and other indicators. But how do acids and bases react together? You will react a number of acids with the base NaOH.

OBJECTIVES

In this experiment you will investigate the reactions of acids and bases.

TECHNIQUES

Correct use of pH paper, indicators, and conductivity meters will be used in this experiment. You will also titrate a base with an acid.

Correct reading of a graduated cylinder will be important to record the exact amount. When reading the volume of a solution in a graduated cylinder, you may read the units the glassware is marked off in, plus one more decimal. For example, if the glassware is marked off in milliliters, you may read to tenths of a milliliter. We are allowed one digit of estimation in significant figures. A volume of a solution is read at the bottom of the meniscus (the curved surface of the solution). A solution for which the bottom of the meniscus is exactly at the 5 mL mark should be recorded as having a volume of 5.0 mL. If the meniscus is just above the 5 mL mark, then you must estimate the last digit. You might record 5.1 mL or 5.2 mL, whichever seems the best estimation for the meniscus as it lies between the 5 mL and 6 mL marks.

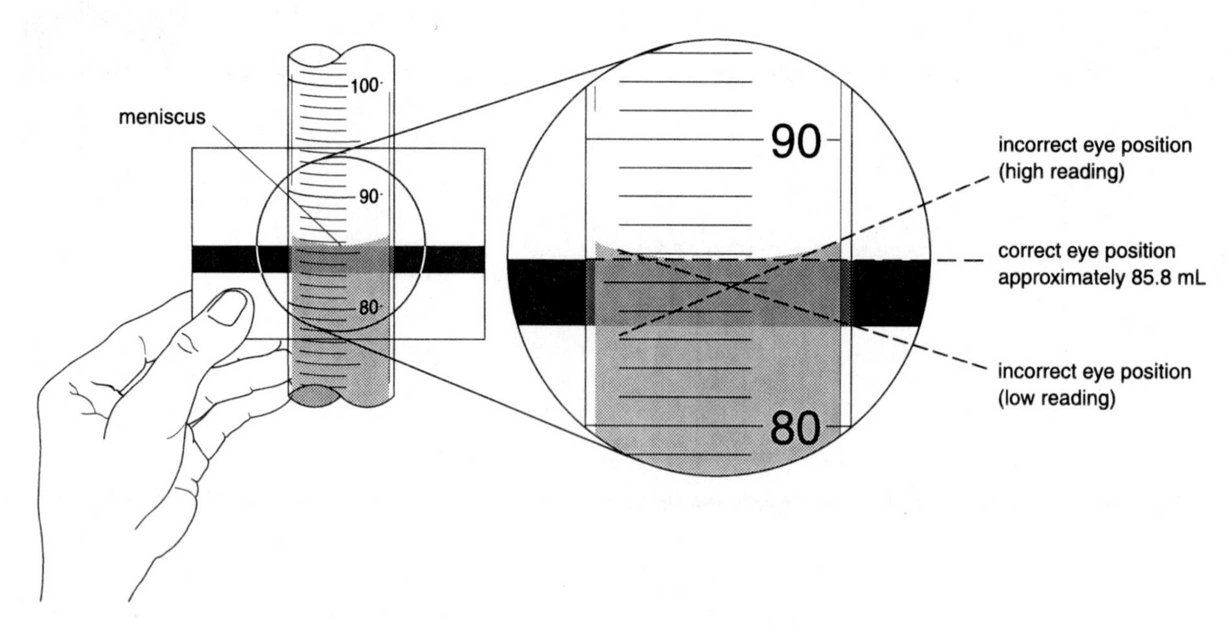

Figure 10A.1 *Correct reading of a graduated cylinder*

APPLICATION PROCEDURES

Part A

1. Test a drop of the 0.010*M* H_2SO_4 on a small piece of pH paper. Record the pH.
2. Test a drop of the 0.10*M* H_2SO_4 on a small piece of pH paper. Record the pH.

Part B

1. Obtain a clean and dry 25-mL or 50-mL graduated cylinder.
2. Add 0.10*M* NaOH to the 5 mL mark. Remember to read the bottom of the meniscus.
3. Add 2 drops of bromothymol blue to this solution and record the color.
4. Titrate this solution with 0.10*M* HCl by adding the HCl dropwise from a dropper bottle and swirling after the addition, until one drop of HCl changes the original color of the solution. If you keep adding acid after this point, the solution will become yellow.
5. When the color changes, record the total volume in the graduated cylinder.
6. Record the approximate volume of 0.10*M* HCl added to obtain the color change. Touch a drop of the solution to a piece of pH paper. Record the pH.
7. Clean and dry the graduated cylinder. Repeat steps 1–6 with 0.10*M* of $HC_2H_3O_2$.
8. Clean and dry the graduated cylinder. Repeat steps 1–6 with 0.10*M* H_2SO_4. Be sure to clean and dry the cylinder before putting the glassware away.

Part C

1. Obtain a conductivity meter. Calibrate the meter as your instructor directs.
2. Obtain a clean and dry 50-mL beaker.
3. Measure 20 mL of 0.010*M* $Ba(OH)_2$ in a graduated cylinder and add it to the beaker.
4. Record the conductivity of the solution.
5. Clean and dry the graduated cylinder. Measure exactly 25 mL of 0.010*M* H_2SO_4 in the graduated cylinder. Use a disposable pipet to get the exact volume.
6. Add a small amount of the acid to the beaker containing $Ba(OH)_2$ with the pipet and swirl the solution, then record the total volume of acid left in the graduated cylinder and the conductivity of the solution.
7. Continue adding the acid until the conductivity is reduced to zero. As the conductivity nears zero, beginning adding the acid dropwise.
8. Test a drop of the solution from the beaker on pH paper. Record the pH.

Name: ______________________ Instructor: ______________________

Date (of Lab Meeting): ______________ Course/Section: ______________

Prelab Exercises

Experiment 10A Reactions of Acids and Bases

1. What is a meniscus?

2. If you are using a ruler that is marked off in centimeters and are measuring the length of a piece of litmus paper, what length should you record for a paper that measures 2/3 the way between 4 and 5 centimeters?

3. What is meant by an acidic solution?

4. Describe what color you expect with bromothymol blue in an acidic solution. (Consult your data from experiment 10.)

5. What is meant by a basic solution?

6. Describe what color you expect with bromothymol blue in a basic solution. (Consult your data from experiment 10.)

7. What is a conductivity meter used for?

Date ____________ Student's Signature ______________________________________

Name: ____________________

Instructor: ____________________ Course/Section: ____________________

Partner's Name: ____________________ Date: ____________________

Report Form

Experiment 10A Reactions of Acids and Bases

Data

Part A

pH of 0.010 M H_2SO_4 __________

pH of 0.10 M H_2SO_4 __________

Part B

Color of solution containing NaOH and 2 drops of bromothymol blue ______________

	Total volume after the addition of acid results in a color change	*Volume of acid added*	*New color of solution*	*pH*
HCl				
$HC_2H_3O_2$				
H_2SO_4				

Part C

Conductivity of 0.010 M $Ba(OH)_2$ ______________

Addition of H_2SO_4	*Total volume of acid remaining in the cylinder*	*Volume of acid added*	*Conductivity (mS)*
1			
2			
3			
4			

Addition of H_2SO_4	*Total volume of acid remaining in the cylinder*	*Volume of acid added*	*Conductivity (mS)*
5			
6			
7			
8			
9			
10			

If more lines are needed, continue in your lab notebook.

pH of solution with a conductivity about zero mS ______________

Date ____________ Instructor's Signature ______________________________

Analysis

1. Which concentration of H_2SO_4 is the most dilute?

2. Which concentration of H_2SO_4 has the most hydrogen ions? Explain your reasoning.

3. Which concentration of H_2SO_4 has the lowest pH?

4. Write the balanced equations for the 3 reactions in part B.

5. Explain why the conductivity in Part C was reduced.

6. Draw a particle view of a container of $Ba(OH)_2$, then draw a particle view of the container after some H_2SO_4 has been added.

Concept Questions

1. Describe the effect concentration has on the pH of acids.

2. Describe the reaction of an acid and a base.

Date ____________ Student's Signature ______________________________

Name: ______________________________

Instructor: ______________________________ Course/Section: ______________

Partner's Name: ______________________________ Date: ______________

Demo Report

Experiment 10A **Reactions of Acids and Bases**

1. Describe demonstration #1 (burning Mg over water).

2. Is the resulting solution acidic, basic, or neutral? Give evidence for your choice.

3. Describe demonstration #2 (goldenrod paper).

4. Is window cleaner acidic, basic, or neutral? Give evidence for your choice.

5. Predict the effect of spraying 0.1 M NaOH on goldenrod paper. Give evidence for your prediction.

6. Predict the effect of spraying 0.1 M HCl on goldenrod paper. Give evidence for your prediction.

7. Describe any other demonstration.

Date ____________ Student's Signature ____

Reactions of Metals and Compounds

INTRODUCTION

Reactions that take place without the continual input of energy are called spontaneous reactions. Reactions that only take place with the continual input of energy are called nonspontaneous reactions. You will be looking at both types of reactions in the experiment.

OBJECTIVES

In this experiment you will investigate reactions between metals and compounds.

TECHNIQUES

Use of well or spot plates helps limit the amount of chemicals needed. You will also use an electroplating technique in Part B.

EXPLORATION PROCEDURES

Part A

1. Obtain a strip of magnesium, zinc, and copper. Gently clean the metals with steel wool or fine sandpaper until they are shiny.
2. Obtain a spot or well place. Add enough 0.1*M* $Zn(NO_3)_2$ to fill 3 wells or spots about 3/4 full.
3. Place one end of each of the metal strips into the 3 wells. Observe the wells for about 5 minutes. Remove each piece of metal and visually inspect it. You may need to gently wipe the metal with a piece of white paper towel to check for any deposits. Record your observations.
4. Gently clean the metals again with steel wool or fine sandpaper until they are shiny again.
5. Add enough 0.1*M* $Cu(NO_3)_2$ to fill 3 wells or spots about 3/4 full.
6. Place the metal strips into the 3 wells containing $Cu(NO_3)_2$. Observe the wells for about 5 minutes. Then remove each piece of metal and make observations. You may need to gently wipe the metal with a

piece of white paper towel to check for any deposits. Record your observations.

7. Gently clean the metals a third time with steel wool or fine sandpaper until they are shiny again.
8. Add enough 0.1*M* $Mg(NO_3)_2$ to fill 3 wells or spots about 3/4 full.
9. Place the metal strips into the 3 wells containing $Mg(NO_3)_2$. Observe the wells for about 5 minutes. Then remove each piece of metal and make observations. You may need to gently wipe the metal with a piece of white paper towel to check for any deposits. Record your observations.
10. Clean your work area as directed by your instructor.

Part B

1. Obtain about 40 mL of 0.1*M* $Zn(NO_3)_2$ in a clean 100 mL beaker, a long strip of zinc, a penny, and a battery.
2. Clean a strip of zinc with steel wool or fine sandpaper until it is shiny. Gently buff a penny with steel wool until it is shiny. (If you are using a "new" penny take care that you don't remove the copper layer.) Once the penny is clean and shiny, only touch the penny by the edges or with a gloved hand.
3. Obtain two pieces of wire that have an alligator clip on each end. Attach an alligator clip of one wire to the strip of zinc and an alligator clip of the other wire to the penny.
4. Put both the strip of zinc and the penny into the beaker containing the $Zn(NO_3)_2$. Do not let the zinc strip and the penny touch.
5. One partner should carefully hold the free alligator clip on the wire attached to the zinc to the positive pole of the battery. The other partner should carefully hold the free alligator clip on the wire attached to the penny to the negative pole of the battery. Continue to hold these connections for 3–4 minutes, observing the beaker, zinc strip, and coin.
6. After breaking the connection, carefully raise the penny and the zinc strip out of the solution. Record your observations.

Name: ______________________ Instructor: ______________________

Date (of Lab Meeting): ______________________ Course/Section: ______________________

Prelab Exercises

Experiment 11 Reactions of Metals and Compounds

1. What are the ions that are in an aqueous solution of $Zn(NO_3)_2$?

2. What are the characteristics of metals?

3. Define what is meant by a compound?

4. What is the difference between Mg metal and Mg ions?

5. Magnesium nitrate contains magnesium metal or magnesium ions. Which choice is the best and explain why.

6. Answer any question provided by your instructor.

Date ____________ Student's Signature ______________________________________

Name: ______________________________

Instructor: ______________________________ Course/Section: ______________

Partner's Name: ______________________________ Date: ______________

Report Form

Experiment 11 Reactions of Metals and Compounds

Data

Part A

	Cu^{2+} *(aq)*	Zn^{2+} *(aq)*	Mg^{2+} *(aq)*
Cu			
Zn			
Mg			

Part B

Sketch the set up of the beaker, zinc strip, wires, and penny.

Observations of the beaker, zinc strip, and penny during and after being connected to the battery.

Date ____________ Instructor's Signature ____________________________________

Analysis

1. Rank the pieces of metal (Cu, Zn, and Mg) from most reactive to least.

2. Write the equations for the reactions of the metals that you observed.

3. What changes occur to the metal reactants in the equations you wrote as they transition to products?

4. Did all the metals react with every solution? If so, what explanation could you give. If not, list the combinations that did not react.

5. Rank the solutions (Cu^{2+}, Zn^{2+}, and Mg^{2+}) from most reactive to least.

6. Write the equations for the reactions of the solutions that you observed.

7. What relationship do you see between the ranking of the metals and the ranking of the solutions?

8. In Part B, you had zinc ions and nitrate ions in the solution. Did a reaction occur? If so, what were the reactants?

9. In Part B, what reaction did you observe? Explain your reasoning.

10. How does the reaction in part B compare with the combinations that did or did not react from part A?

11. Did the reaction in part B occur when the leads were NOT attached to the battery? Would you say that the reaction in part B was spontaneous? Explain.

Concept Questions

1. When a compound and a metal react, explain how they react.

2. If a compound and a metal do not react spontaneously, how could the reaction be forced?

Application Questions

1. When copper metal and a silver nitrate solution are combined, the copper metal will begin to be coated with a shiny deposit. Does a reaction occur? If so, is the reaction spontaneous? If not, what could be supplied to cause the reaction to happen.

2. When silver metal and a copper nitrate solution are combined, the solution and metal remains unchanged. Does a reaction occur? If so, is the reaction spontaneous? If not, what could be supplied to cause the reaction to happen.

3. When nickel metal and a copper nitrate solution are combined, nothing happens. When a battery is attached to the nickel and to a strip of copper that are submerged in the solution, the nickel starts to get a reddish shiny deposit. Does a reaction occur? If so, is the reaction spontaneous? If not, what could be supplied to cause the reaction to happen.

4. If potassium is more reactive than magnesium, would you expect potassium to react with magnesium nitrate? Explain your response.

Date ____________ Student's Signature ______________________________________

Name: ______________________________

Instructor: ______________________________ Course/Section: ______________

Partner's Name: ______________________________ Date: ______________________

Demo Report

Experiment 11 **Reactions of Metals and Compounds**

1. Describe demonstration #1.

2. Can you explain the results you observed?

3. Describe any other demonstration.

Date ____________ Student's Signature ______________________________

Reaction Energetics

INTRODUCTION

One of the first batteries was made by alternating layers of zinc, paper that had been soaked in salt water, and silver. Coins are made of metals. Pennies (after 1984) are made of zinc with a thin copper coating, nickels are made of an alloy of copper and nickel, dimes and quarters are made of an alloy of copper and nickel over a copper core. You know that batteries supply electrical energy. You can make a battery made of coins.

OBJECTIVES

In this experiment you will investigate the combination of coins that will give you a battery with the largest voltage.

TECHNIQUES

You will use a multimeter to detect any voltages produced. The exact procedure will depend on the brand of multimeter that you have. Your instructor will provide this information. You will also prepare saturated solution. Saturated solutions contain all of the solute that will dissolve. You know that you have a saturated solution when a small amount of the solute remains undissolved at the bottom of the container even after you have stirred the solution.

EXPLORATION PROCEDURES

Part A

1. In a clean 100-mL beaker, prepare about 20 mL of saturated saltwater solution.
2. Tear a piece of filter paper into pieces that are slightly larger than the pennies, nickels, dimes, and quarters. Soak these in the saltwater solution.
3. Clean a collection of pennies, nickels, dimes, and quarters by placing them in a 3*M* acetic acid solution and/or cleaning them with steel wool. Vinegar is made of acetic acid.

4. Obtain a paper clip. Straighten the outer loop and twist it so that it sticks up when the paper clip is flat on the table. If the paper clip becomes corroded, you will need to obtain a new paper clip and straighten it.
5. Place a nickel on top of the paper clip, followed by a piece of filter paper that has been soaked in saltwater. The paper should just barely extend over the edges.
6. Next place a penny on top of the filter paper, followed by another nickel and another saltwater-soaked piece of filter paper. Make sure the two pieces of filter paper aren't touching. Finally, place a penny on top of the last piece of paper.
7. Connect one lead of the multimeter to the end of the paper clip that is sticking up and one lead on top of the pile. Turn on the multimeter to the settings provided by your instructor. Measure the voltage of the pile.

Part B

1. You and your partner should design an experiment to investigate the effects of varying the height of the pile, the effects of using different types of coins, and the effects of using different order in the arrangement of coins and paper. The design is your prelab and must include:

   Problem Statement: This includes a few sentences describing the specific question(s) you are trying to answer with your experiment.

   Proposed Procedures: This section contains the materials and equipment that you will use, the type of data you will collect (the variables you will measure), and the number of trials you are proposing. Remember to discuss safety considerations. Your planned experimentation should take up at least 2/3 of the allotted time. Remember that multiple runs will establish the reliability of your data.
2. After your instructor has given your design approval for safety and chemical concerns, conduct your experiment.
3. After completing the data collection, you will write a lab report. Although you collect data and share ideas with a partner, you will be expected to write the final lab report independently. Your grade will depend on the thoroughness of your investigation, the presentation of your data, the careful analysis of the data, and the logic put in to give reasonable results and explanations.
4. The lab report MUST include the following four sections:

   Problem Statement: This includes a few sentences describing what specific questions you were trying to answer with your experiment.

   Procedures: This section contains the materials and equipment that you actually used (these may differ from those you proposed), the type of data collected (the variables measured), and the number of trials done. Remember to discuss safety considerations. Your experimentation should take up at least 2/3 of the time allotted.

Data/Analysis: Include the data you collected. Data should be in tables when possible, with easy to read labels. Analysis of the data should also be included (analysis = what does your data tell you). Graphs (with labels, units, and titles) and any mathematical relationships should be given, and the connection to the data should be shown.

Conclusion: This is the generalization or explanation you have deduced from your experiment. This is also the place to make explanations for any data results that are counter to logical chemical ideas and to describe how you would change the experiment if you repeated it. Be sure to give your reply and evidence for these three questions:

Did the height of the pile make a difference in the voltage?

What types of coins yielded the most voltage?

What arrangement of coins and paper yielded the most voltage?

Name: ______________________ Instructor: ______________________

Date (of Lab Meeting): ______________________ Course/Section: ______________________

Prelab Exercises

Experiment 12 Reaction Energetics

1. Define what is meant by saturated saltwater solution. How would you prepare this solution?

2. Problem Statement for Part B.

3. Proposed Procedures (use the back and insert more sheets if needed)

Date ____________ Student's Signature ____________________________________

Name: ______________________________

Instructor: ______________________________ Course/Section: ______________

Partner's Name: ______________________________ Date: ______________

Report Form

Experiment 12 **Reaction Energetics**

Data

Voltage from the arrangement of nickel-paper-penny-nickel-paper-penny __________V

Remaining data is collected in your notebook.

Date ____________ Instructor's Signature ______________________________

Lab Report

Attach this sheet to your lab report that includes the **Problem Statement** (actual), **Procedures** (actual), **Data/Analysis**, and **Conclusion**.

Concept Questions

1. What problems did you encounter in making your battery?

2. Explain how you overcame these problems?

Date ____________ Student's Signature ______________________________

Name: ______________________________

Instructor: ______________________________ Course/Section: ______________

Partner's Name: ______________________________ Date: ______________

Demo Report

Experiment 12 Reaction Energetics

1. Describe demonstration #1

2. Can you explain the results you observed?

3. Describe demonstration #2.

4. Can you explain the results you observed?

5. Describe any other demonstration.

Date ____________ Student's Signature ______________________________________

Water Testing

INTRODUCTION

Water is one of the most common substances on Earth. However, the quality of water can widely vary. You will investigate a number of attributes of a water sample. The pH is a measure of the hydrogen ion content and defines acids and bases. The total hardness of water was once thought of as the capacity of water to precipitate soap. Soap is precipitated chiefly by the calcium and magnesium ions present. Total hardness is a measure of the amount of calcium and magnesium ions and is important if water softening is considered. The total alkalinity is the amount of bases in water (usually as bicarbonate or carbonate ions). Total alkalinity is related to pH and corrosion.

Chlorine is a disinfectant for water. The free chlorine amount indicates the combined concentrations of HOCl, OCl^-, and Cl_2. Total chlorine is the sum of free and combined chlorine (chlorine that is in combination with ammonia or organic compounds). Free chlorine is a stronger disinfectant than combined chlorine, but some combined chlorine compounds are more stable and can provide disinfection over long exposure periods than free chlorine.

Many of the measures of water quality are often done in units of parts per million. This is equilavent to milligrams per liter.

OBJECTIVES

In this experiment you will investigate the quality of a water sample and compare it to distilled water

TECHNIQUES

Correct use of pH paper, test paper, and conductivity meters will be used in this experiment.

EXPLORATION PROCEDURES

Part A

1. Obtain a conductivity meter. Calibrate the meter as your instructor directs.

2. Obtain a clean and dry 50-mL beaker.
3. Measure about 40 mL of a water sample in the beaker.
4. Record the conductivity of the solution.
5. Using a stirring rod, transfer a drop of the solution to a piece of pH paper. Record the pH.
6. Dip one test strip into the water for about 5 seconds while gently moving the strip back and forth in the sample.
7. Remove the strip and shake it over a paper towel to remove excess water
8. Wait about 25 seconds, then using the color charts match each of the 5 sections and record pH, total hardness, total alkalinity, free chlorine, and total chlorine.

Part B

1. Obtain a conductivity meter. Calibrate the meter as your instructor directs.
2. Obtain a clean and dry 50-mL beaker.
3. Measure about 40 mL of distilled water in the beaker.
4. Record the conductivity of the solution.
5. Using a stirring rod, transfer a drop of the solution to a piece of pH paper. Record the pH.
6. Dip one test strip into the water for about 5 seconds while gently moving the strip back and forth in the sample.
7. Remove the strip and shake it over a paper towel to remove excess water
8. Wait about 25 seconds, then using the color charts match each of the 5 sections and record pH, total hardness, total alkalinity, free chlorine, and total chlorine.

Name: ______________________ Instructor: ______________________

Date (of Lab Meeting): ______________ Course/Section: ______________

Prelab Exercises

Experiment 13 Water Testing

1. Define what is meant by the total hardness of water.

2. Define what is meant by the total alkalinity of water.

3. Define what is meant by the total chlorine of water

4. What is meant by an acidic solution? Describe what pH you expect for an acidic solution. (Consult your data from experiment 10.)

5. What is meant by a basic solution? Describe what pH you expect for a basic solution. (Consult your data from experiment 10.)

6. What is a conductivity meter used for?

7. Answer any question provided by your instructor.

Date ____________ Student's Signature ____

Name: ______________________________

Instructor: ______________________________ Course/Section: ______________

Partner's Name: ______________________________ Date: ______________

Report Form

Experiment 13 Water Testing

Data

	Water Sample	Distilled Water
Conductivity		
pH from pH paper		
pH		
Total hardness		
Total alkalinity		
Free chlorine		
Total chlorine		

Date ____________ Instructor's Signature ______________________________

Analysis

1. Compare the pH measured with pH paper and the pH from the test strip. Explain any similarities and differences.

2. What relationship should there be between total alkalinity and pH? Does your data show this relationship?

3. From the comparison between the free and total chlorine, what can you say about the amount of combined chlorine in your water sample? ...in the distilled water?

Concept Questions

1. Which tests depend on ions present in the water?

2. What similarities are there between your water sample and distilled water?

3. What differences are there between your water sample and distilled water?

4. Draw a particle view of your water sample.

Application Questions

1. If the conductivity of a water sample is 0.5 mS, what would values would you expect for:

Total hardness	*Total alkalinity*	*Free chlorine*	*Combined chlorine*	*pH*

Explain your predictions

2. If the conductivity of a water sample is zero, what would values would you expect for:

Total hardness	*Total alkalinity*	*Free chlorine*	*Combined chlorine*	*pH*

Explain your predictions

Date ____________ Student's Signature ____________________________

Name: ______________________________

Instructor: ______________________________ Course/Section: ______________

Partner's Name: ______________________________ Date: ______________

Demo Report

Experiment 13 **Water Testing**

1. Describe demonstration #1 (sodium acetate).

2. Can you explain the results you observed?

3. Describe any other demonstration.

Date __________ Student's Signature ______________________________

Fuels

INTRODUCTION

Heat is a useful form of energy that is a product of combustion (burning). The heat evolved during combustion reactions can be used to do work such as drive an electrical generator or a motor. Most of the common fuels contain carbon and hydrogen. The products then include carbon dioxide and water. Below are some examples of combustion reactions involving compounds that contain carbon and hydrogen.

The burning of natural gas:

$$CH_4(g) + 2\ O_2(g) \rightarrow CO_2(g) + 2\ H_2O(g) + \text{heat}$$

The burning of wood alcohol (methanol):

$$2\ CH_3OH(g) + 3\ O_2(g) \rightarrow 2\ CO_2(g) + 4\ H_2O(g) + \text{heat}$$

The burning of wood:

$$(C_6(H_2O)_5)_x(s) + 6X\ O_2(g) \rightarrow 6X\ CO_2(g) + 5X\ H_2O(g) + \text{heat}$$

$$(x = 900 \text{ to } 6000)$$

It takes 1 calorie (cal) of heat to raise the temperature of one gram of water one degree Celsius. This relationship is also called the specific heat of water. Joules and BTUs are sometimes used as the units for heat but for this experiment we will only use calories (cal). The dietary or food calorie (C) is equal to 1,000 small calories (cal). A candy bar contains 250 C (food calories), which is 250,000 calories.

In this experiment you will heat a sample of water with the heat evolved from combustion reactions similar to the reactions above. Is there any relationship in the makeup of the fuel and the heat evolved? If so, what factors affect this relationship? If the mass of water and its temperature change are known, one can calculate the heat content change of a sample of water using the formula:

$$\text{Heat evolved} = \text{mass of water} \times \text{specific heat of water} \times \text{change in temperature of the water}$$

$$= \text{mass of water} \times 1\ (\text{cal/gram}\cdot{}^\circ\text{C}) \times \Delta T$$

OBJECTIVES

You will investigate the combustion of a number of substances. Some of those substances may include: ethanol or grain alcohol (CH_3CH_2OH), propyl alcohol or propanol ($CH_3CH_2CH_2OH$), butyl alcohol or butanol ($CH_3CH_2CH_2CH_2OH$), lamp oil or kerosene ($C_{10}H_{22}$), candle wax or paraffin ($C_{26}H_{54}$). Then you will use the results to look for patterns that involve mass burned, rate of heat transfer, the calories transferred and structure of the fuel.

TECHNIQUES

You will construct the apparatus shown below. It is constructed from:

1 Ring stand and iron ring

2. Soft drink (soda) can supported in the iron ring with a glass rod

3. Alcohol burner containing a fuel as designated by your instructor

4. Alcohol thermometer that will also be used to carefully stir the water that will be placed in the soda can once you reach that point in the experiment

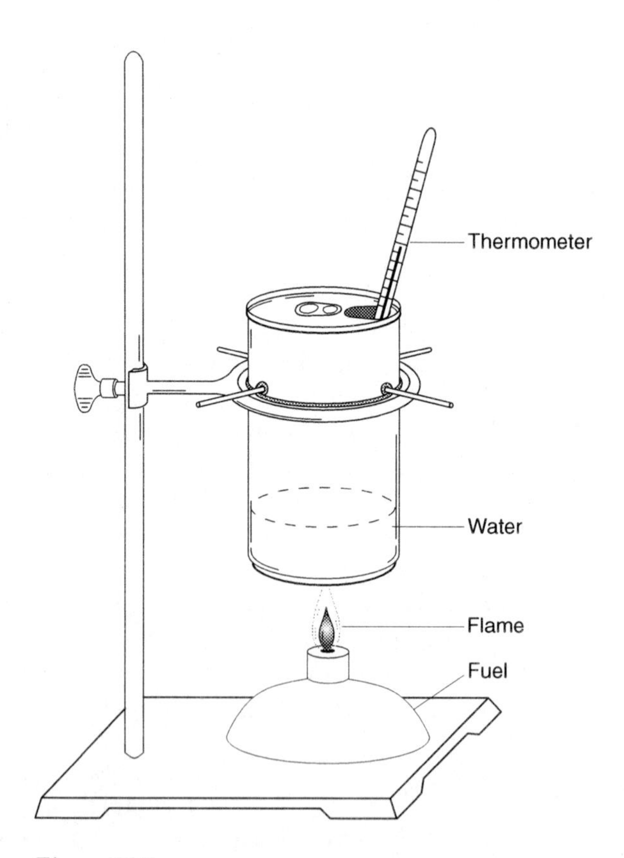

Figure 14.1

EXPLORATION PROCEDURES

1. Lift the can from the iron ring. Remove the glass rod and thermometer. Empty the soda can if it is not already empty.
2. Take the empty can to the top loader balance and determine its mass to the nearest 0.01 gram. Record the value.
3. Pour about 100 mL of distilled water into the can. Reweigh the can containing the water. Record the value.
4. Return the glass rod to the holes in the can and place the can in the iron ring.
5. Your instructor will assign you one of the possible fuels. Take an alcohol burner containing the assigned fuel to the balance and determine its mass to the nearest 0.01 grams. Record the value.
6. Place the thermometer in the water in the can. Allow about 2 minutes for the thermometer to reach the temperature of the water. Record the temperature of the water to the nearest 0.01°C. If your thermometer is not divided into marks of 0.1°C, estimate the reading to the nearest 0.1°C.
7. Light the alcohol burner and adjust the height of the can so that the flame just touches the bottom of the can. Record the time.
8. Continue to heat the can for exactly 5 minutes.
9. Turn off the burner, stir the water and record the maximum temperature that it reached and the time at which that temperature is reached.
10. Repeat steps 1 through 9 in order to obtain a second set of data.
11. Return the burner to its designated place and obtain a candle and a square of heavy paper.
12. Use a small ball of clay to position the candle on the square of paper if it is not already attached to the paper.
13. Repeat steps 1-10 but this time you are using the candle in place of the alcohol burner. (You should have a total of 4 runs.)
14. Clean and return all equipment to its proper location.

Name: ______________________ Instructor: ______________________

Date (of Lab Meeting): ______________________ Course/Section: ______________________

Prelab Exercises

Experiment 14 **Fuels**

1. What is a calorie?

2. What is the specific heat of water?

3. Do you expect the transfer of heat from the reaction to the water to be highly efficient? (Will all of the heat from the reaction be transferred to the water?)

4. Do you expect the efficiency in question #3 above to be consistent from one experimental run to the next?

5. Draw a picture of the apparatus that will be used to heat water with a candle instead of the alcohol burner. Label the (1) soda can, (2) glass rod, (3) ring stand, (4) iron ring, (5) water, (6) thermometer, (7) candle, and (8) paper or cardboard onto which the candle is secured.

Date ____________ Student's Signature ______________________________________

Name: ______________________________

Instructor: ______________________________ Course/Section: ______________

Partner's Name: ______________________________ Date: ______________

Report Form

Experiment 14 **Fuels**

Data

Run/Fuel	*Mass of empty can*	*Mass of can plus water*	*Mass of alcohol burner (before):*	*Mass of alcohol burner (after):*	*Initial temperature of the water*	*Final temperature of the water*
RUN #1 *Fuel:* ________						
RUN #2 *Fuel:* ________						
RUN #3 *Candle*						
RUN #4 *Candle*						

Date ____________ Instructor's Signature ______________________________

Analysis

Run/Fuel	*Mass of water in the system*	*Temperature change of the water*	*Mass of fuel consumed*	*Number of calories gained by the water*
RUN #1 *Fuel:* ________				
RUN #2 *Fuel:* ________				
AVERAGE *For RUNS #1 and #2*				

CLASS AVERAGE for same fuel (Identity of fuel used: ________________________)

Average of runs #1 and #2: calories = ___________, mass = _________

Run/Fuel	*Mass of water in the system*	*Temperature change of the water*	*Mass of fuel consumed*	*Number of calories gained by the water*
RUN #3 *Candle*				
RUN #4 *Candle*				
AVERAGE *For RUNS #3 and #4*				

CLASS AVERAGE for the candle

Average: calories = __________, mass = ________

CLASS AVERAGE for a different fuel (Identity of fuel used: ________________________)

Average: calories = __________, mass = ________

CLASS AVERAGE for a different fuel (Identity of fuel used: ________________________)

Average: calories = __________, mass = ________

CLASS AVERAGE for a different fuel (Identity of fuel used: ________________________)

Average: calories = __________, mass = ________

Concept Questions

1. What was the energy per gram of fuel burned for each fuel used in the class?

2. Which fuel yielded the most energy per gram of fuel burned?

3. Which fuel yielded the most energy in 5 minutes? Which fuel yielded the most energy per minute?

4. Which alcohol yielded the most energy per gram of alcohol? The least?

5. Compare your answer to the last question to ratio of carbon atoms to oxygen atom in the formulas of the two alcohols.

Application Questions

1. When might one select a fuel based upon the energy delivered per unit of time?

2. When might one select a fuel based upon the energy delivered per unit of mass (or volume)?

3. What energy per gram of alcohol would you expect for hexanol ($CH_3CH_2CH_2CH_2CH_2CH_2OH$)? Explain your prediction.

Date ____________ Student's Signature ______________________________

Name: ______________________________

Instructor: ______________________________ Course/Section: ______________

Partner's Name: ______________________________ Date: ______________

Demo Report

Experiment 14 Fuels

1. Describe the portion of a flame that is the hottest.

2. Why are alcohol flames dangerous?

3. Describe demonstration #2.

4. Describe demonstration #3, if applicable.

5. Describe demonstration #4, if applicable.

Date ____________ Student's Signature ______________________________________

The Nature of Polymers

INTRODUCTION

Polymers may be natural or synthetic: for example, wood or plastic. White glue contains long strands of the polymer (polyvinyl acetate) suspended in water. Borax can act as a *cross-linker* that will connect the polyvinyl acetate strands at various locations to make new, even larger polymeric strands called GLUEP. What are polymers? What properties do they exhibit? Can the properties be changed? When certain characteristics are desired in the final product, chemists combine substances in a variety of ways to make a variety of new products one of which may have the desired properties.

OBJECTIVES

In this experiment you are to conduct investigations into the differences in polymers prepared by slightly different procedures.

REVIEW CONCEPTS

This experiment uses the concepts of synthetic organic polymers.

REVIEW TECHNIQUES

Measuring volume, transfer of solutions, and making observations

CAUTION

It is recommended that you dispose of GLUEP in a waste can. Do not put GLUEP down the drain; it can clog the drain. GLUEP should be stored in a sealed plastic bag. It should be refrigerated to retard mold growth. GLUEP will leave a water mark on wood furniture and can stick to carpet. Vinegar will de-gel the GLUEP from carpet, which then can be washed with soap and water.

EXPLORATION PROCEDURES

1. Use glue, water and borax in 4 different proportions to make GLUEP of different types. Your instructor will assign which mixture(s) you and your partner should make.

	Water	*Glue*	*Borax*
Mixture A	30 mL	15 mL	10 mL
Mixture B	15 mL	15 mL	10 mL
Mixture C	15 mL	15 mL	20 mL
Mixture D	0 mL	15 mL	10 mL

2. Measure out the desired amount of glue and of water.
3. Mix the water and glue in a container. You may add food coloring at this point.
4. Measure the desired amount of 4% borax solution into a clean graduated cylinder.
5. One member of the team is to stir the glue and water mixture with the craft stick while the other pours in the borax solution.
6. When the GLUEP becomes a gel, remove it from the container. Discard the remaining liquid. Knead the GLUEP like dough for a few minutes.
7. Does the consistency change if the GLUEP is left undisturbed for a while? Does the consistency change with stirring or agitation?
8. Record the following for one of the products.
   a. Describe the color and texture of the GLUEP A, B, C, or D.
   b. Describe its odor and consistency.
   c. Record its mass.
   d. Form a ball from the material. Poke the glob quickly with your finger. Now poke it slowly. Which way does your finger go in further?
   e. Put the material in the palm of your hand. Close your fist quickly around it, and then close it slowly. What differences do you observe?
   f. Form it into a ball by rolling it on the tabletop. Hold the ball in your open palm. Does it retain its shape?
   g. Roll it on a flat surface. Does the ball bounce? Does it follow a straight smooth path or does it bounce as it rolls?

h. Form it into a rope by rolling it on the tabletop. Pull it apart slowly; then pull it apart quickly.

i. Pat it into a pancake. Make an imprint with a small object like a coin or key and observe.

j. Pat it into a thin sheet. Hold it up by one edge and make observations.

9. Repeat #8 for each mix until you have observed Mixtures A, B, C, and D.

10. Record all data for use in answering questions in the Report Form.

Name: ______________________ Instructor: ______________________

Date (of Lab Meeting): ______________________ Course/Section: ______________________

Prelab Exercises

Experiment 15 The Nature of Polymers

1. Describe how you measure milliliters.

2. How does your textbook describe polymers?

3. Describe the role of a cross linker.

4. Give the formula of vinyl acetate (consult your textbook).

5. Give a representation of the general reaction that can be used to represent the polymerization of vinyl acetate to form poly(vinyl acetate).

6. Give two uses of poly(vinyl acetate).

7. Answer other questions assigned by your instructor.

Date ____________ Student's Signature ______________________________________

Name: ______________________________

Instructor: ______________________________ Course/Section: ______________

Partner's Name: ______________________________ Date: ______________

Report Form

Experiment 15 The Nature of Polymers

Data

Describe each mixture:

Mixture	*Color*	*Texture*	*Odor*	*Consistency*	*Elasticity*	*Mass*
A						
B						
C						
D						

Record any additional observations

Record the results for A, B, C and D when you:

Mixture	*A*	*B*	*C*	*D*
Poking slowly then quickly with your finger				
Closing fist quickly then slowly				
Holding in your open palm				
Bouncing and rolling				
Pulling apart slowly and quickly				
After imprinting a small object				
Holding a thin sheet from the top end				

Date ____________ Instructor's Signature ________________________________

Analysis

1. Compare the unmixed glue, water, and borax with the mixture of the three substances (the GLUEP).

2. What were the variables in the directions for making each type of GUEP?

3. Which variable was changed in each mix (A, B, C, and D)?

4. How does the consistency of the GLUEP change when you change the proportions of the variables in the mixture?

5. Does any mixture (A, B, C, or D) stick to your hands when you touch it?

6. Which mixture (A, B, C, or D) yields a GLUEP that has the greatest tendency to return to its natural shape? Explain your reasoning.

7. Which mixture (A, B, C, or D) yields a GLUEP that is least likely to return to its natural shape? Explain your reasoning.

8. As you rolled the GLUEP, did you observe if the heat from your hands affected its consistency?

9. Which mixture did you personally like the best? Why?

10. Which form of GLUEP must have the greater amount of water trapped within the cross-linked polymer? Why did you pick that one?

Concept Questions

1. Would you classify a polymer as a solid, a liquid, a gas, or a combination? Explain your answer.

2. How do you explain the different properties you found in the mixtures?

3. What is the relationship between the properties exhibited and the contents of the mixtures for A, B, C, or D?

4. Suggest some possible uses of GLUEP mixtures A, B, C, and D.

Date ____________ Student's Signature ______________________________________

Name: ______________________________

Instructor: ______________________________ Course/Section: ______________

Partner's Name: ______________________________ Date: ______________

Demo Report

Experiment 15 **The Nature of Polymers**

Nylon

1. How many carbons were in the diacid and in the diamine used in 1928 by Wallace Carthers to make the first nylon?

2. What diacid was used in the demo in which you observed nylon being formed?

3. What diamine was used in the demo in which you observed nylon being formed?

4. Is nylon an addition polymer or a condensation polymer?

5. What properties of nylon can you observe without touching it?

Polyvinyl Acohol

1. What is the purpose of the borax in this demonstration?

2. Describe what you observed during this demonstration.

3. What properties of polyvinyl alcohol can you observe without touching it?

Other Demos (if any)

1. Describe what you observed during these demonstrations.

2. Answer other questions assigned by your instructor.

Date ____________ Student's Signature ______________________________

Separating Plastics

INTRODUCTION

In the previous experiments, you investigated the properties of polymers and the relationship between mass and volume. In this experiment, you will be given a mixture of plastics. Plastics are polymers and have different densities, which tells us about their composition. We recycle plastics by their densities, which are designated by a number.

Number	*Name*	*Abbreviation*
#1	Polyethylene terephthalate	PET
#2	High-density polyethylene	HDPE
#3	Polyvinyl chloride	PVC
#4	Low-density polyethylene	LDPE
#5	Polypropylene	PP
#6	Polystyrene	PS
#7	Other (can be mixtures and composites)	

The following liquids and solutions have different densities:

Vegetable Oil	Density = 0.92 g/mL
1:1 Ethanol/H_2O	Density = 0.94 g/mL
Water	Density = 1.0 g/mL
10 % NaCl (*aq*)	Density = 1.08 g/mL

OBJECTIVES

In this experiment you will be given a number of samples of plastics. You must determine the densities of each type of plastic and devise a method to identify an unknown plastic or to separate a group of plastic samples. In this experiment you will design your own procedures.

TECHNIQUES

Observation and inference are used in this experiment.

APPLICATION PROCEDURES

1. The design is your prelab. The design must include:

   Problem Statement: This includes a few sentences describing the specific question(s) you are trying to answer with your experiment.

   Proposed Procedures: This section contains the materials and equipment that you will use, the type of data you will collect (the variables you will measure), and the number of trials you are proposing. Remember to discuss safety considerations. Your planned experimentation should take up 2/3 of the lab period. Remember that multiple runs will establish the reliability of your data. Try to keep the amounts small and the concentrations dilute.

2. After your instructor has given your design approval for safety and chemical concerns. Conduct your experiment.

3. After completing the data collection, you will write a lab report. Although you collect data and share ideas with a partner, you will be expected to write the final lab report independently. Your grade will depend on the thoroughness of your investigation, the presentation of your data, the careful analysis of the data, and the logic put in to give reasonable results and explanations.

4. The lab report MUST include the following four sections:

   Problem Statement: This includes a few sentences describing what specific questions you were trying to answer with your experiment.

   Procedures: This section contains the materials and equipment that you actually used (these may differ from those you proposed), the type of data collected (the variables measured), and the number of trials done. Remember to discuss safety considerations. Your experimentation should take up 2/3 of the lab period

   Data/Analysis: Include the data you collected. Data should be in tables when possible, with easy to read labels. Analysis of the data should also be included (analysis = what does your data tell you). Graphs (with labels, units, and titles), mathematical relationships, chemical equations, and algebraic equations should be given, and the connection to the data should be shown.

Conclusion: This is the generalization or explanation you have deduced from your experiment. This is also the place to make explanations for any data results that are counter to logical chemical ideas and to describe how you would change the experiment if you repeated it.

Name: ______________________ Instructor: ______________________

Date (of Lab Meeting): ______________________ Course/Section: ______________________

Prelab Exercises

Experiment 15A **Separating Plastics**

1. Problem Statement

2. Proposed Procedures (insert more sheets if needed)

Instructor's Approval and Comments:

Name: ______________________________

Instructor: ______________________________ Course/Section: ______________

Partner's Name: ______________________________ Date: ______________

Report Form

Experiment 15A **Separating Plastics**

Data

Data is collected in your notebook.

Date __________ Instructor's Signature ______________________________

Lab Report

Attach this sheet to your lab report that includes the Problem Statement (actual), Procedures (actual), Data/Analysis, and Conclusion.

Concept Questions

1. What are the most important factors to keep in mind when designing an experiment? Discuss at least three.

2. Describe the densities of plastics #1 through #7.

Less than 0.92	
Less than 0.94	
Less than 1.0	
Less than 1.08	
More than 1.08	

Density Ranking

(lowest) ______ ______ ______ ______ ______ ______ (highest)

3. What can you propose about the arrangement of materials in plastic #2 and #4?

Date ____________ Student's Signature ______________________________________

Name: ______________________________

Instructor: ______________________________ Course/Section: ______________

Partner's Name: ______________________________ Date: ______________

Demo Report

Experiment 15A Separating Plastics

Polyurethane Foam

1. Describe what you observed during this demonstration.

2. What properties of polyurethane can you observe without touching it?

3. Explain the changes in density that you observed.

Other Demos (if any)

1. Describe what you observed during these demonstrations.

2. Answer other questions assigned by your instructor.

Date ____________ Student's Signature ______________________________________